AIRCRAFT WOODWORK

Ruth Spencer

After 14 years — nothing new. Ruth

New York
MODERN AIRCRAFT SERIES
A Division of Sports Car Press

Published in New York by Sports Car Press, Ltd. and
simultaneously in Ontario, Canada, by General Publishing Co.

Manufactured in the United States of America.

Library of Congress Catalog Card No. 77-162586
ISBN: 0-87112-052-6
Crown Item Number: 509016

2nd Printing, July, 1976

MODERN AIRCRAFT SERIES

Edited by Joe Christy

A new series of popular-priced books on aircraft and their operation for everyone interested in privately owned planes. Each volume is written by an expert in the field, and is printed on fine, white paper and profusely illustrated with photographs and diagrams.

Learning to Fly

Your Pilot's License
Legal Guide for Pilots & Owners
Your Jet Pilot Rating
Pilot's Digest of FAA Regulations

Navigation and Weather

Nav/Com Guide for Pilots
Cockpit Navigation Guide
Pilot's Weather Guide
Air Traffic Control
Instrument Flying Guide
Computer Guide
Practical Guide to Aviation Weather

Flying Sports

Skysurfing
Parachuting for Sport
Soaring Guide
Modern Aerobatics
Club Flying

Vocations and Avocations

Agricultural Aviation
Taking Pictures from the Air
How to Draw Airplanes
Jet Aircraft Engines
Lightplane Vacationing

Individual Light Planes

The Single-Engine Cessnas
The Piper Cub Story
The Single-Engine Beechcrafts
All About Helicopters
The Single-Engine Piper
Buying a Used Plane

Classics and Classes

Classic Biplanes
Classic Military Biplanes
Classic Monoplanes

Military

Aces & Planes of WWI
The Curtiss Hawk Fighters
Aircraft Armament

For the Homebuilder

Guide to Homebuilts
Aircraft Dope & Fabric
Aircraft Woodwork
Lightplane Construction & Repair
Custom Lightplane Design
Modern Lightplane Engines
Restoration of Antique & Classic Planes
Guide to Homebuilt Rotorcraft

Published by
SPORTS CAR PRESS

Distributed by Crown Publishers
1 Park Avenue, New York, N.Y. 10016

Contents

page

1 Wood, Wind and Sky 7

2 Materials and Equipment 9

3 Construction: Fuselage and Tail 23

4 The Wings 33

5 Construction of Spars 41

6 Construction of Ribs 47

7 Making Bows 63

8 Wing Assembly 71

9 Repair of Wood Airframes 93

10 Repair of Wings 99

1. Wood, Wind and Sky

From dugout canoe to wooden sailing ship to the racing sloop on the waterways; the wheel, the carriage, the automobile on land; the simplest of gliders to World War I fighting planes, the airplanes of the 1920's and '30's, and now the supersonic jets in our fast shrinking air envelope today. The rate of change is almost too rapid for one individual to keep up with, and sport planes are the only types he can reasonably build for himself. He may well be a part of the new technology but the materials and methods of mass production are usually beyond his personal grasp. He is not competing with production aircraft; he is satisfying a very personal desire to have and fly a particular kind of airplane. Often, what he learns in building his own plane is of value in his profession, whatever it may be. There is no limit to the imagination and ingenuity of man.

Wood seems archaic to those who have grown up with metal aircraft construction, and for modern commercial and military planes it would be. Wood construction is not feasible for mass production because its nature requires hand working. It is amazing that wood planes were produced in such large numbers during World War I and for many years afterward. The German metal Junkers was an interesting curiousity for many years while even Tony Fokker stayed with wood. The Fokker tri-motor as well as the single engine planes had the all-wood, full cantilever wing, `concurrently with the all-metal Ford tri-motor. By World War II the composite airframe was in common use with metal tube airframes and all or part wood wings, in the training planes. The more advanced training planes in the U.S. were all metal.

The ubiquitous Stearman primary training plane later became the backbone of the Agricultural Aircraft industry in this country contributing much to the high production rates of modern farming in this country and many others. The deHavilland Moths performed similar services in all British oriented parts of the world. In both cases this was due primarily to the large numbers of these aircraft available as well as their general adaptability and repairability. Both kinds could be endlessly repaired and rebuilt. When an aluminum structure is badly bent, it has "had it".

The wood wing has some characteristics which will always be superior for many special purposes and these will be discussed as we go on. The use of wood by amateur builders will remain popular for very practical reasons, too. It is to this latter group, along with aircraft mechanics who have not had occasion to become familiar with aircraft woodwork, that this book is primarily addressed.

Anyone building or repairing wood aircraft structures should have on hand for close study and reference, the FAA publication: *AC Acceptable Methods, Techniques and Practices Aircraft Inspection and Repair Manual 43.13-1.* Its older equivalent, *Manual 18,* is identical insofar as *Woodwork, Dope,* and *Fabric* are concerned.

Chapter I of *43.13-1* states the requirements and limits for wood construction and repairs. Because the mechanic or builder should have and use this Manual, the information therein will *not* be repeated here. The purpose of this book is to be of practical help in doing the work, with advisory notes on points to be watched for while using elementary equipment.

One does not have to be a professional woodworker or cabinet maker to work on airplanes. A basic knowledge of the use of common tools, the ability to read plans and directions, and a good sense for observing characteristics of wood is sufficient. Most of the homebuilt wood airplanes I have seen have demonstrated a high degree of skill and integrity in construction even though their owner-builders worked in totally different professions. Patience combined with genuine interest and a reasonable degree of manual skill are the chief requirements. A basic knowledge of aerodynamics and experience in the maintenance of airplanes of any type helps a lot.

2. Materials & Equipment

We hear and read of other kinds of wood being used as a substitute for the long-time standard: Sitka Spruce. A chart on page 2 of *Manual 43.13-1* shows the properties of other woods *acceptable* for aircraft use, as compared with spruce. While some are superior in some ways, it is clear that spruce has the best all around characteristics for aircraft structures. The *Uniformity* of Aircraft Grade *(inspected)* Sitka Spruce is characteristic of other woods which may in some particular ways be superior. The toughness, uniformity, and flexibility of spruce is is far superior. It is nice wood to work with; it does not gum up as does pine in cutting tools or sandpaper.

Please Note: Not all Sitka Spruce is of aircraft grade. Local lumber merchants do not carry it. The spruce they may have in stock is intended for building purposes, ladders, etc. They may occasionally have a few pieces with very good grain which would pass inspection for grain direction, spacing, and absence of defects and which, if they passed as to moisture content, would be useable. However, if moisture content is unknown considerable warping could and probably would develop. I once purchased a supply of rough spruce, most of which was all right (it was sold to me as "aircraft spruce"), but one board bowed so badly edgewise that it was unuseable for anything but corner-blocks and other such occasional needs in places requiring only short sections. That was the last spruce purchased from any dubious source. When you order *Certified, Inspected, Aircraft Grade Spruce* it will carry the Inspector's stamp from the mill. You will get your money's worth and there will be no waste because of poor wood. It is kiln dried; it will not warp except under conditions of improper storage or use. It should be stored flat on a level surface, raised from a concrete floor with cross-strips so moisture from the concrete will not wet the wood. Of course this is true for all wood. Do not stand boards on end to store because this will cause them to bow. Keep wood clean, dry, and neatly stacked so edges will not be damaged. Buying rough wood is a gamble, regardless of source. Even the supplier's inspectors pass only "surfaced" pieces so both sides and edges can be inspected for defects. (See page 3, *AC 43.13-1*)

Since 1909 the prime supplier of Certified Aircraft Grade Spruce in the United States has been the Posey Manufacturing Company of Hoquiam, Washington. Their principal business is making sounding boards for pianos and other musical instruments where the requirement is for uniformity of wood grain and density without defects which would reduce or distort the resonate qualities of pitch and depth of tone. *Aircraft Spruce* is rather a by-product; the "deluxe" pieces of spruce reaching the mill on Gray's Harbor. The choice wood used to come down from Alaska, but much of that crop is presently being shipped to Japan, and most of the Posey material is now coming out of British Columbia. Before reaching Hoquiam by rail or truck, it has already been sorted two or three times, rough, and cut into cants of varying lengths up to 24 ft. to 26 ft. by 6 in. x 6 in. to 8 in. x 8 in. These are then sawed into boards dimensions varied to take best advantage of grain direction and eliminate defects that by then may be apparent.

The boards are kiln dried for seven to eight days in Posey's scrap-wood stoked, steam heated kiln which is the size of a large room. The layers of spruce boards are separated by cross-wise strips of wood so the heat will be evenly distributed. The moisture content is reduced to 10% or 12%. Spruce intended for aircraft use stays in the kiln about a day longer than that for instrument sounding boards.

Unfortunately, the rate of growth of the best 100-year-old Sitka Spruce trees, which grow to a diameter of six to eight feet, has not been able to keep pace with the demand. Vast quantity were used for airplane wings during the second World War. During a one month period in 1943, Posey shipped 23 carloads to Wichita for Stearman wings. They also furnished laminated spars for the large transport gliders built at that time. The Alaska spruce is slower growing and consequently of finer and more even grain than the British Columbia spruce.

Pieces over 1⅝″ thickness must be specially selected and kiln dried, and it may take considerable time for suitable pieces to come to the mill after Posey sends along the order to various shipping points up in the north woods. Lengths available without grain run-out

After close scrutiny for any defects the Posey Mfg. Co. Inspector puts his stamp of approval on both sides of every piece of *Certificated Aircraft Spruce* before it is wrapped for shipment.

exceeding limits usually are available in 16 ft. to 18 ft. pieces. Spars of extra large depth dimension can be edge-glued with proper equipment. An example of this is the main spar of the Davis D-1 parasol monoplane of the 1930's. The thick laminated spar of the famous Bellanca Cruisair and the Viking requires extra wide spruce for depth at the root end.

Many planes of WW I and the 1920's had wide (fore and aft) but shallow (vertically) spar sections. The deHavilland Gypsy Moth (DH 60-M) shown has wing spars and ribs typical of this period. The airfoil is shallow with under-camber and the spars are wide to provide strength. They are deeply routed, ⅜" to lighten them. The Meyers OTW, built some twenty years later, has spars of similar dimensions but they are not routed; they are very heavy but no stronger. The very solid spar of this cross section lacks flexibility. The airplane is considerably heavier than the very light Gypsy Moth. It is much easier to obtain spruce spar stock for the thinner and deeper

Originally an all-wood airframe, the fuselage was changed to square steel tube for 1928 and '29 models by De Havilland and Wright (in U.S.A.), but the rest of the Gypsy Moth and the later Tiger Moths remained wood. Dudley R. Kelly's complete restoration is ready for a flight at the 1967 AAA Fly-In at Ottumwa, Iowa. Even the cockpit coaming is all-wood.

The author's home-built *Termite* fills basement work area during pre-cover assembly. Some builders have made do in less space than this. At this stage the most common question: "How are you going to get it out?" Many modifications were made from the beginning on this one.

spar than for the old style shallow, wide ones. This has to do with cutting for vertical grain direction, necessary for spar strength.

Since we cannot wait for giant spruce trees to grow, other styles of spar construction, discussed in a later chapter, can be utilized. Those building small to average sized planes with conventional airfoils and spars will have no trouble getting the spruce needed for them, but for certain of the "antique" types one may be faced with a real problem when trying to replace spars "as per original." One suggested solution: *laminate* for required width.

There are now very few of the old all-wood fuselage frames left. A relatively modern example of the monocoque type is the beautifully restored Lockheed Vega, N105W an exact copy of Wiley Post's famous plane. This is an excellent example of a high performance, all-wood, cantilever wing and tail aircraft. The entire plane is plywood skinned and fabric covered for a strong structure and sleek finish. A number of Vegas set substantial records in the 1930's, for both speed and distance. Many of the fastest racing places of the 1930's were all-wood construction. The large number of wood airplanes which were very successful will not be enumerated but we wish to point out that the best reason for not continuing to produce them was *not* lack of performance or durability but because metal is the material of mass production methods. For custom-built planes, the material used depends entirely on the kind of flying expected of the airplane and the personal capabilities of the builder.

Primarily, at the present time, aircraft of wood construction in entirety or in part are of the sport type with the notable exception of the excellently-built production Bellancas, four-place craft for general aviation use. Some high performance soaring gliders are all or part wood; some are all-metal and some new ones are now being built of light weight plastic shells of very clean aerodynamic design. Materials are very much the choice of the builder. Wood spars are desirable in aerobatic airplanes because they combine flexibility with strength. There are no rivets to work loose in the endless and varied flexing of aerobatic maneuvers. The same is true of the long cantilever wings of the soaring gliders which can be laminated and tapered for great strength and flexibility. For lightness, these spars may be of "box" construction, instead of being laminated. Some of the small home-built planes also use the box-spar which will be discussed later.

Any discussion of wood in aircraft use must include propellers. It was interesting to learn a couple years ago, when visiting at the Sensenich Propeller Corporation, that they are now manufacturing *more* wood props than for many long years. This is evidently because of the increasing number of light planes being built and restored which use the smaller engines, as well as the lower purchase price compared to metal props. The wood propeller is lighter in weight though no one argues the better efficiency of the metal fixed pitch propeller. But for a light plane used solely for the fun of flying, speed is not the most important thing and the wood prop, correctly matched to engine and plane, gets the job done very well. Sensenich has a very competent engineer to advise on the question of just *what is* the best propeller installation for a particular aircraft. This can be a fascinating subject. A tour of the Sensenich plant is very interesting, both in the metal prop department and in the wood shop, and here again, wood propellers are second in production to hardwood table tops and butcher blocks.

New propellers start out as roughly contoured, flat pieces, very carefully laminated so that the total is in balance, even at this stage. They are glued in a fast-setting electronic "oven", machine contoured, and then hand-finished before varnishing. Old props are totally reconditioned. The metal and fabric tipping is removed, rivet holes inspected (not reused if enlarged; they are plugged and new holes are made) and any damaged areas are repaired, often with wood inlays. The blades are retipped, the prop is re-balanced and varnished. Except for the wood being a little darker, the reconditioned propeller looks like new. Of course, they recieve some propellers which cannot be restored to airworthy condition. Sensenich can reproduce any propeller they have ever made.

There are constantly new developments in engineering propellers for various applications in order to achieve the greatest efficiency, with changes made in length, pitch, and blade style. The problems range from desired speed, rate-of-climb, rpm (too much or too little) to sound (noise). One peculiarly-shaped blade, quite wide near the hub end and tapered at the tip, stands in the large room containing the shapers and other special equipment. I was told it was one of six blades being made for Lockheed; beyond that its purpose was not known (by my guide). A couple years later I recognized these blades in a photo of one of the "Quiet" planes developed by Lockheed for the U. S. Military services as an observation plane in South East Asia.

Custom propeller makers are busy: Ole Fahlin, Ray Hegy, and others. The lively interest in "original" airplanes being built, with modifications and restorations of old ones, is going to keep the wood prop-makers going for a long time. It is hoped that they are training a few young men to carry on in the coming years.

In consideration of the overall cost of even the most elementary small airplane, there is little reason to skimp on materials simply save a few dollars in the purchase of either spruce or plywood. Strangely enough, the less there is to building the plane, the greater the temptation to use second rate materials by many people who try to justify this by convincing themselves that the cheaper materials are "just as good." In the case of the amateur-builder, maybe it is — for him. I want the best I can get. I have seen enough second-rate materials and workmanship when repairing and rebuilding wings to be convinced that only accepted materials are worth bothering with. The texture and uniformity of non-aircraft plywoods do not compare favorably with approved aircraft quality, either mahogany or birch; they do not hold up as well. Materials specifications are not just fantasies; they are the results of many years of trial and usage or many times, trial and error. There is room for argument regarding a few details in the *FAA Manual 18* or the later *AC 43.13-1,* but for the most part they are based on experience gained from many years of service and maintenance.

It is true that many old-time airplanes were built of materials no longer recommended. It is also true that most of these planes are long gone. This should tell you something. It is a lot cheaper to learn from the experience of others. A few dollars more at time of construction can save a lot of grief and money later. It will unquestionably give you a much finer airplane, with greater sale value, if that concerns you.

One of the most common faults of amateur aircraft designers and builders is to build *too heavy.* By using the correct kind of materials in the best way much weight can be saved by not adding more "meat" to beef it up when using poorer grades of wood. The greater the weight the greater the wing loading and the poorer the performance.

Build well and keep it light.

One of the attractions of working with wood is that it can be done in places and under conditions that would not be possible for other methods of construction. There have been a couple of instances when I have repaired or built wing panels in the living room when the regular work area was occupied by other component parts. However, the shop equipment was readily accessible; it just required a lot of extra running back and forth. In places like this it is necessary to spread a large heavy canvas over the floor to protect it from drops of glue and varnish and scratches. It can be done.

The minimum space required must be sufficient to provide a couple feet clearance around the wing or fuselage being worked on with additional room for a six foot work bench (a good, solid one), your power equipment and hand tools. Being crowded is a handicap but at the same time it tends to force one to keep things neat and organized. In cold weather the room should be heated to a comfortable 72F° minimum, because glues, like Butyrate dope, should be used at this minimum temperature. It should be reasonably dry also; too high humidity is detrimental for proper setting of glue.

It is advisable to have an exhaust type window fan (as for dope work) to remove sawdust flung about by the sander, especially that from plywood. The bonding glues used in plywood are highly irritating to the respiratory system.

Build a pair of sturdy sawhorses of convenient working height for yourself. If too low, you will very soon have an aching back; if too high, you will not be able to reach nor see well enough. A second pair is a great convenience because you will need them for cutting from full sheets of plywood, lengths of spruce, and endless other uses. A piece of plywood layed on them serves nicely as a second work bench, or a place to stack the multitude of parts and pieces for a new set of ribs, etc. Be certain each *pair* of sawhorses match in height.

An uneven floor is a nuisance. The sawhorses, and particularly your power tools, must have a solid, even footing. If the floor is not even and level it will be necessary to shimm under the legs so they will be firm and level.

Unless you intend to do an extensive amount of wood work, that is, more than one little airplane, furniture and household carpentry, it is less costly and certainly faster to hire mill work done by a good mill or cabinet shop where a skilled machine operator will do a precise job for you in less time than you could think about how to go about it. His hourly rate may sound high, but this is less costly and more satisfactory than to buy the machines or take the time to set up makeshift arrangements. Those with the know-how can set up power tools with very little outlay of money, but these folks are in the minority. Generally, it costs *more* in both time and money to try to get by cheaply than to pay a fair price to have one or two special milling jobs done (accurately) for you. This is particularly true when your wood supplies must be ordered and shipped from a long distance, with substantial cost and some waiting time. The equipment you have will determine just how much you can do yourself. The mill work mentioned here can be pretty well predicted and much of this will depend on how you order it.

As previously stated, do not order rough spruce, but material "surfaced both sides", whatever the use may be. Spar stock (so stated in your order) should be ordered to your finished dimensions; it will be shipped 1/16″ over-size, edgewise, to allow for bevelling. It is shipped square, not bevelled; there is too much chance of corners being damaged in shipping and besides, that would be additional cost for labor. Rib stock is best ordered as *flat* stock, surfaced both sides, in random widths and lengths (state minimums), shipped in bundles. This is ordered to the depth (vertically) of the cap-strips, giving you vertical grain (for strength) for rib construction. You will have this ripped lengthwise to the *width* required on table-saw or band-saw. This is one of the jobs best hired done for uniformity and accuracy. Unless you have a power planer, the bevelling of the spars is also best done quickly and accurately professionally. The airframe longerons and truss members are also ripped from boards ordered to thickness, also surfaced two sides. If you are going to laminate wood wing tip bows, spruce (or ash) can be supplied in ⅛″ thickness, and again, this will be ripped to provide the maximum width needed (at the leading edge of the wing). From here on, you will be on your own

A length of spruce is clamped to the band-saw table as a fence for cutting ¼″sq. rib material. The two cut edges were then finished on the disk-sander. Machines run at high speed for wood; slow for cutting steel; this includes drills.

to cut and finish incidental parts and pieces, of which there will be a multitude.

It is assumed that since you are building or restoring an airplane you have the standard tools to work on one: screwdrivers, wrenches, cutters, etc. I will be blunt. If you do not have them, find out what is needed and buy them. Anyone who cannot buy the tools to work on his own airplane should not have one, or should take it to a shop. Neither a mechanic working for his living nor another airplane owner can afford to lend out tools, and there is nothing he hates more than to be asked. An occasional special tool, possibly, but unless you really know how to use it explain just what it is that you need to have done and let him do it for you with his own tools rather than ask to borrow. If you do borrow, return it clean and in good condition — promptly. Tools are increasingly expensive and some are hard to get. So do unto others as you would have them do unto you. Do not be ashamed to admit you do not know how to use a special piece of equipment. If you rent equipment, have it checked out by someone who fully understands its use. Both spruce and aircraft plywood rapidly dull cutting tools; keep them sharp. Welding and metal working tools necessary for steel tube and aluminum work are in a different department. The following are essential for aircraft woodwork. Substitution of hand tools for power tools will be noted with an asterisk. A skilled woodworker might use some alternate equipment, but the tools listed have served me for a long time while working on a variety of more or less conventional airplanes. One more *special note:* Keep oils and greases away from the work area, bench, sawhorses, and all tools. The wood used in construction will readily absorb anything fluid or greasy. Work with clean hands and do not allow anyone to smear greasy, dirty hands, or anything else, on the wood or on the work bench being used. Use paper towels as wipers when gluing.

Hand Tools

A medium size and solid bench *Vise.* Use thin blocks of wood between the jaws when holding parts for an airframe; *don't crush.*

C-Clamps; approximately a dozen large and a dozen small.

Parallel Clamps; a couple pairs.

Tack-Hammer with magnetized head.

Sheet-Metal Hammer for aluminum parts incidental to the work. No "all-wood" airplane ever really is!

Bucking-Bar, partially rounded one side only.
Hand-Saws; 1 cross-cut and 1 small back-blade.
Plane; small to medium size.
Knife; (pocket) for cutting and scraping.
Wood Chisel; 1" wide.
File (rasp); very coarse one side, medium on other.
Brace & Bit; for larger holes.

Power Tools

Drill Motor or *Hand Drill;* for smaller holes. These to be used only when a *drill-press* is not available or cannot be used. do NOT make holes in spars with other than a drill-press, *or* any large holes that require the use of a *fly-cutter.* When using a *fly-cutter,* go part way through from one side, then reverse and complete cutting the hole from the opposite side. ALWAYS have a piece of wood *under* the work when drilling holes through the wood, whether spruce or plywood. Stands are available for securing an electric drill-motor to the work bench for use as substitute for a drill-press. These are not as good as the drill-press, but far better than the hand-held drill motor.

*Drill-Press** (See above note)

Drill Bits; for the appropriate kinds of drills listed. I have found the very simple flat, blade-type drills with the cutting edges on the face of the diameter, to be very good for both wood and aluminum. They make a good clean hole in either material. Use standard twist drills for the smaller sizes. Be certain of accurate size. Keep them sharp.

Band-Saw (12"); invaluable for nearly everything. Use a ½" wide band for cutting heavy material *(wood only)* and a narrow one, 3/16" or ¼", for all other cutting. The table can be tilted for sawing any desired angle (bevel). This is true also of the sander. This saw is a "must" for good aircraft woodwork.

*Table-Saw**; essential if you intend to rip your own cap-strip, etc., but *expensive*. For limited use, I recommend the alternative of having the mill work done at a cabinet shop or mill.

Disc Sander; essential for all fitting, removing burrs from cap-strip, sanding stacks of gussets for uniformity and shape, etc.

*Router**; for use with a pattern to cut out plywood ribs and to cut lightening holes in plywood ribs or formers. Plywood ribs can be cut out on a band-saw and lightening holes can be cut out with a fly-cutter or disk type hole-cutter.

*Planer** attachment to be used with above *router;* fine for cutting bevels on spars. For limited use, not essential.

Routing of spars is precision work which is best done by a skilled operator. I would not recommend attempting it with a portable machine.

Standard Measuring Tools

6 ft. (min.) *Tape Rule*

6 in. and 12 in. Scales, in 1/32″ increments minimum, 1/64″ preferred.

Square and *Protractor*

*Set of Trammell Points** and bar for squaring wings during assembly. See text for alternative.

Other Materials

Sanding Paper; actually, medium and fine grit *garnet* paper or aluminum oxide paper, for all wood sanding. Use a medium grit disk also on the disk-sander; again the Garnet or Aluminum is preferred.

*Weldwood Plastic Resin Glue** (recommended for all aircraft woodwork), or other *approved glue* (Refer *AC 43.13-1*). Use cold water for mixing. Mix in small cup for use within three hours.

Marine Spar Varnish or other highly weather-resistant finish which will penetrate the fibres of the wood and be durable and also compatible with the dope or other fabric finish to be used.

There are at present some synthetic finishes which may prove superior to the Marine Spar Varnish but at this time, I hesitate to recommend them, pending trial. They are more expensive than the Marine Spar Varnish, but whether they are also better remains to be seen. This is true likewise of various adhesives. The (above) Weldwood Glue is excellent and extremely durable; I have never had a glue joint fail with it. It is less expensive and much easier to use (neatly) than many other kinds.

Never mix fresh glue in a cup containing a wet residue. One with a dried and hardened residue can be re-used without it ever flaking into a fresh batch. But a still soft mixture will break up into the new mix and spoil it as would any other foreign matter. If water is put into a little plastic cup after pouring out un-used glue (Weldwood), it can be completely cleaned out before preparing a fresh batch of glue as the glue does not bond permanently to the smooth plastic surface of the cup. I previously used little paper cups for glue and re-used them endlessly, just cleaning out old (wet) glue as much as possible then letting the rest harden before using that cup again; rotating a few in this way was no problem.

3. Construction: Fuselage and Tail

There are four basic types of fuselage construction, with countless variations, dating back to the earliest flying machines, evolving rapidly during the first World War. During that period the Pratt truss was the most common. The longerons with vertical and horizontal braces were held together with metal corner fittings screwed or bolted in place and all diagonal interbracing accomplished by wires (or cables) and turn-buckles. This was a holdover from the days of non-covered fuselage frames. The style extended into the 1920's. It now represents a very nostalgic era of aviation — and it worked.

The Warren truss utilizes diagonal bracing of approximately the same material and dimensions as the longerons and other primary frame members, normally secured by glued and nailed gussets at all joints. This is also our familiar welded steel tube airframe. With but few exceptions, this was a development following the war. Many wood, Warren truss airframes also had plywood (glued and nailed) outer sheathing which contributed greatly to their strength. The early deHavilland 60-G Gypsy Moth, like the D.H.-4, is a well known example. This was later built as the 60-M model with square steel tube frame of the same dimensions; the wings and tail surfaces were still of wood. The writer's homebuilt pictured is a typical simple Warren truss fuselage, the easiest for an amateur to build.

The World War I Pfalz D-XII fighter was one of the most interesting and advanced airframes at that time and since. Molded plywood "shells" were secured on bulkheads, with the strips of plywood winding in a spiral fashion around the frame for an aerodynamically clean and strong fuseage. In recent years, similar methods take advantage of plastics and much more will be done along this line in the future. The adaptability of both plastics and aluminum out-dated the all-wood monocoque type of construction, and again, the Lockheed Vegas and Orions come to mind as the best examples which are reasonably familiar because of the wonderful records they set for all the world to shoot at during the 1930's.

The aluminum fuselage was the next logical step. One (of only three) *metal* fuselage Vega is currently owned by Robert L. Taylor, founder and head of the Antique Airplane Association. The heavy gauge aluminum of the fuselage skin of this airplane is notable; there is nothing flimsy about it. The wood wing is plywood covered, like the earlier model, as are the tail surfaces. NC12288 has been nicely restored with very handsome leather covered seats and upholstery in the cabin. The plywood skinned wing and tail are of course fabric covered for a clean and durable finish. The Lindberghs' Lockheed Sirius, of "North to the Orient" fame, was an open-cockpit (2-place), low wing version very similar in construction to the better known Vega. Both were fine airplanes which presaged things to come.

Some very interesting original amateur-built designs are now flying which reflect variations and modifications of the above basic styles and there will be countless more in the years ahead. "Where there's a will, there's a way" is as true now as ever. And as long as there are trees, man will use them to build with.

The actual techniques of fuselage construction do not differ from those used in building other parts, such as wings, so what is learned of all following details applies as well to one as the other. The fundamentals are common to all components just as the techniques of welding are applied to all parts of a welded airframe. Each set of plans for the various amateur-built airplanes carries details peculiar to that one design and no one guide book can anticipate them all, but there are certain points to be watched for that are common to all with wood construction.

The simple Warren truss fuselage frame is the easiest to build and it is actually just like a balsa model plane. A jig is made for the side frames on two or more (as needed) lengths of common plywood that is solid enough (⅝" minimum) to maintain a flat and level surface. Hold the plywood together, end to end, with sturdy lengths of pine nailed or screwed to both plywood sections so they can not shift. Unless you make a very long table with saw-horses and boards, this can be worked on the floor. A table would provide a more comfortable working height. Lay out the basic frame plan in pencil, directly on the plywood and make the jig to accurately locate every piece in the side frames. Use blocks that are a little less thick than the frame members so the glue squeezed out at the joints will not hold the frame to the jig at time of removal. It is best not to put these jig blocks at the joints, but an inch or so away at each piece. This will give clearance to pry the frame from the jig if necessary after it has been assembled with its gussets.

Keep in mind that the one jig will serve to build both right and left side frames, so include, and identify, all details for both sides. If there is to be plywood sheathing from the fire-wall to perhaps aft of the cockpit, put that on one side while in the jig then put the gussets on the reverse side after it is out of the jig. The opposite side should be assembled first, however, so after it is removed from the jig, and before the sheathing is glued and nailed on, use it for a pattern to mark out where the frame members will be for nailing. Trace around the frame for both right and left plywood side sheathing. The marked side of both pieces to be facing outward so you can see where to nail. (The glue will be put on the frame). It is plain that however you make the one, the other will be assembled in reverse order so you will not wind up with two right sides or two left sides. The visible frame plan traced onto the outsides of the plywood sheathing will also come in handy later on in construction for installation of controls and other parts to be mounted at frame members and it is very important to know exactly where the structural supports are. It will save much time and probable errors.

Many of the small all-wood home-built planes have straight top longerons, from fire-wall to tail-post, so it is easiest and best to construct the fuselage by assembling the side frames inverted, with the top longerons secured in a second jig laid out for the top longerons and cross members. Make this jig for the top plan form as you did for the side frames, to hold them vertically (bottoms up) while joining them with the cross members with gussets made to fit the joints. By firmly bracing the up-standing frames, the bottom cross members can be glued and nailed in place.

It is essential that this assembly be kept perfectly in line, from center-line at the fire-wall to the tail-post. Be very careful when bringing the side frames together at the tail-post. It may require extra large, bar-type clamps to bring it together uniformly at each station. The precise technique will be determined by the details of style and dimensions of the structural material. My own very typical fuselage is of ⅞″ square spruce (this is quite heavy) with birch plywood gussets and sheathing from fire-wall to aft of the cockpit (sides and on bottom) and the first station aft of the fire-wall is an extra strong "sandwich", with ½" plywood framing glued on both the front and rear faces. Main components attach at this station: the front wing cabanes at the top corners; the main landing gear legs and front lift struts at the lower corners. There is no taper from cockpit, forward to the fire-wall, so I assembled only the square forward part of this fuselage in the jig, then pulled the side frames together from aft of the cockpit to the tail-post. I think the use of a full length jig would be the better way, however. It was necessary to cut matching bevels on each individual tail-post, which had been built into the two side frames, so they could be accurately mated for gluing together. This little detail had not been taken into account in the plans and such omissions had to be figured out as construction proceded. There were very sturdy gussets on both top and bottom which has proved to be very strong. The unintentional "laminated" tail-post is sturdy, if not quite according to plan. So we learn from experience; this is the kind of thing to try to anticipate by carefully studying the drawings in detail before cutting the first stick of wood or piece of tubing.

The better plans for amateur-built planes will generally advise on construction procedures and sequences for the easiest and best method of fabrication and assembly with details applying to special points. Often there is only one sequence to follow in assembling a unit and it may be entirely different than here described. The author has not designed airplanes, just built and repaired them.

Abrupt bends in wood are not desireable and should be avoided, so do not be mislead by drawings that show them this way, such as upward from bottom to tail-post or inward to tail-post from aft of the cockpit. More gradual bends, both ways, in the longerons hold far more strength. The more abrupt and sharper the angle of bend in any wood, the weaker the area involved. For the normal bends in longerons the spruce is flexible enough that there is no need to wet or steam the wood and the gussets connecting the longerons to the cross members will hold them.

Construction of monocoque fuselages by amateur builders is certainly not impossible but it would be more difficult. An internal type jig would be needed to support the fuselage bulkheads while the structural, 45° plywood skin sections were secured after

Termite, **'Broom-Stick" Special, showing many structural details. The baggage compartment "box" contributes to strength of the cockpit area. Toe-hold for getting in and out was idea borrowed from WW I aircraft. A-65-8 engine.**

"Broom-Stick" four years, four landing roll-out bashes (and repairs) and 90-plus flight hours later. *Wood* **damage always negligible; the 4130 steel fittings were the victims. All well now. Neither very fast nor aerobatic, but a lot of fun to fly, and that was the purpose. Covered with Grade-A Cotton and Butyrate Dope. 300 hp Lycoming Stearman alongside owned by W.E. Spencer.**

they were formed with the help of steam. This last detail would probably present the most difficult problem. 45° plywood is used in all applications requiring compound curves; it costs approximately half again as much as standard 90° plywood. All plywood skin, stressed or otherwise, is glued on with the use of nailing strips in order to get a more uniform pressure along all the scarfed edges (over structural members, of course). The nails are removed along with the nailing strips, then the edges sanded smooth with the surface. Any openings for controls or for inspection must be carefully located so as not to weaken the structure, and they will be internally reinforced with plywood, well glued in position. Any lock-nuts or other internal fasteners will be secured to the internal reinforcing before it is glued in place, and the cover for the opening should be made a flush fit with the surrounding skin. Aircraft built in this fashion are generally of the high performance type, and much care goes into such details.

For an amateur-built plane it is much simpler to build the fuselage frame as described, then put a plywood skin around added formers to provide a nicely rounded contour. A compromise and style of construction in its own right, actually, is the so-called "basket-weave", with thin and rather narrow strips of spruce spiralling around the basic frame from nose to tail to provide a light and strong airframe with pleasingly rounded contours. It is reminiscent of the method of construction of the WW I Pfalz which had steamed strips of plywood wound around the fuselage, nose to tail in spiral form, glued and nailed in position to the basic fuselage frame. It was handsome and strong. A modern and increasingly used version is the preformed plastic shell. Only time will prove or disprove the durability of such structural plastics combined with engine vibration and the poundings of use during the service life of an airplane. There is undoubtedly a great future for these materials and much research is being done along these lines, especially in production light planes. Smoother running engines will be a big step forward in developing new building materials for aircraft. The ability of wood structures to absorb vibation and shock is one of its most attractive features and its absorption of noise is notable. My own little plane has spectacularly demonstrated the ability of spruce to bend without breaking while 4130 steel (flat type) fittings twisted like pretzels and had to be replaced. A standard aluminum plane would have been "totalled."

Whatever the style of construction, the important points are well bonded *glue joints* for strength and *adequate drainage*. There should be clear, ¼″ (minimum) drain holes in all bottom-side gussets and skinned areas, with holes and sea-plane type grommets in the fabric covering at all low points.

Every bit of the framework and skin should be smoothly sanded to remove any roughness. The wood should, of course, have been smoothly surfaced before assembly; all edges and corners should be lightly sanded; this is not only neater, it helps prevent minor splintering which could become major. Remove all saw-dust before hand-brushing fresh Marine Spar Varnish. The first coat will be almost entirely absorbed. The second coat will give a gloss; if not, give it a third. This will seal the wood and give good moisture protection. If there is roughness after the first coat of varnish, sand that area smooth before applying the second coat of varnish. The smoother the surface, the less dirt and moisture will be held and the longer your airplane will last.

For upholstering cockpit or cabin interiors it is best to glue and nail-on additional nailing strips to attach it rather than to tack it directly to frame members. The fewer the nail, screw and bolt holes in any part, the better. The appearance of clean, varnished wood is itself very pleasing, inside or out, and I personally prefer this to upholstery in a small plane.

All controls and fittings must be installed with bolts (not screws), nuts (self-locking or cottered) and *wide-area washers* in all wood structures. Secondary items may utilize wood screws, but don't depend too much on them nor use too many of them. Drill (undersize) for screws so the wood will not be split. Drill to accurate size for all bolts. *Do not over-tighten bolts or screws.* Remember that wood is easily crushed.

Seat belts and shoulder harness attachments must be anchored at very strong points of the fuselage frame which will provide maximum resistance to failure in any type of accident; *not* where the frame could collapse inward, releasing tension on the belts. Never attach seat belts or shoulder harness at any center point of a cross member, but at very-well-braced corners and in such a way that a pull would be in line with the longerons rather than across an adjoining member. The seat belt attachment can usually be tied in with the rear landing gear attach fitting for strength and convenience of location relative to the seat. This deserves serious

Aft fuselage portion of Cvjetkovic 2-place CA-65 shows precise workmanship of William Slater. Cabin area has plywood skin both inside and out. Fuselage sides will have full length stringers added. Plane will be powered by Continental 0-200 engine, and have full instruments and electrical system.

planning and an understanding of the forces involved in various types of accidents.

If your plans do not include details for intermediate fuselage formers or the originals have been lost for an antique restoration, a string can be stretched from the most forward to the most rear one to give a straight line for reference to make cardboard patterns for new ones at each station and the stringers can be laid out at the same time. If wood stringers are being used, cut the notches for them to equal depth in each former to maintain straight lines. To keep the outline of the formers from showing, they can be scalloped between stringers, or else cut shallower notches for the stringers so they will not sit as deeply into the formers. However, the first method is better because it will give full sidewise support to the stringers.

On a flat-sided fuselage, the stringer may just be glued and nailed to the sides and bottom with added corner-blocks for additional support. Many steel tube fuselages have brackets for standing off the stringers for the desired contour, and the wood stringers are in this case usually safetied in place with wire. Stringers should be no less than ¼″ x ¾″ on small planes, and larger on big fuselages where there may be considerable distance between supports. If formers are close together the stringers can be of the smaller dimensions. The closer the supports, the better, to prevent them from warping.

William Slater built the turtle-deck for his 2-place Cvjetkovic (see photos) to be removable as a unit for access to the aft portion of the fuselage at some future time should he decide to add equipment or make any changes in that area. This is easily done by using screwed-on angle brackets for attachment. Very small 90° brackets are sufficient. The stringers are glued and nailed to the formers and there are enough of them to make a very solid but light framework.

Any and all wood fairings for contour or other uses, inside or out, must be tightly secured in place with clamps, brackets or tape, and *ALL* must be well varnished for the durability of the wood so they will hold shape and position and so they will not absorb moisture which would deteriorate the adjacent wood or metal structure nor the fabric covering.

Many details in the cockpit area can very well be made of spruce and plywood. Note the photo of the D.H. Gypsy Moth. The cockpit coaming, instrument panels and the baggage compartment are built integrally with the turtle-deck. Incidently, there is nothing prettier than a mahogany instrument panel finished with clear varnish.

Remember that sheets of plywood are heavy; it is very easy to let weight accumulate in this and that detail in far greater degree than one would expect. For instance, stringers used with formers (with cut-out lightening holes) are much lighter than plywood skin. However, do not sacrifice strength for weight saved, particularly in structural areas. Plywood less than 1/16″ thick is like light cardboard and will not hold shape. One sixteenth of an inch is used for gussets in wings and ⅛″ for gussets in the fuselage in most instances. Because of weight, plywood in the cockpit area is generally ⅛″ and sheathing aft of the cockpit is 1/16″. Three sixteenths of an inch is needed for floor-boards and wing-walks but is much too heavy for skin; it is generally enough also for formers and bulkheads in small planes but ¼" is needed in larger planes. One quarter of an inch is best for seats and seat backs because it is stronger and more solid and will be more durable. Use lightening holes with discretion; spruce strips for stiffeners may be advisable to prevent bending or warping. Lighter plywood may be used in this way while retaining the necessary stiffness.

Do not be tempted to "beef-up" unnecessarily. If your plane has been well engineered, leave it alone. Some home-builts are designed too heavy and improvements may be in order. But unless you have considerable experience, get the advice of someone knowledgeable before making changes that will alter the weight and balance of your airplane, whether new or old. In fact, any changes in an ATC'd airplane must be FAA approved. A great many amateur-built planes finish up tail heavy and more often than not, because of numerous "amateur" changes made during construction. Never omit structural pieces; they are there for some good reason.

The cost of a wood airframe, using only first grade materials, is going to be less than steel tube. It will probably also be heavier. *More* flat sheet 4130 steel is required for fittings than in a tube frame and it is amazing how much of this does go into an "all-wood" airplane. Then there are the lift or inter-plane struts, landing gear, cowling, coaming, controls, etc. The average small home-built will generally run from $1500.00 on up, depending on the engine installation, equipment, interior finish, etc. An electrical system adds a lot of weight but it is certainly desireable for extensive cross-country flying. The wood airplane can be just as nicely finished and equipped as any other and it will definitely be quieter because of sound absorption of the wood airframe. Given good care and maintenance, it will last a long time.

4. The Wings

The type or style of the wing is determined by the kind of airplane and its use, considering speed, maneuverability and the load to be carried. The design of every airplane has always been a compromise, with the ultimate aim being the lightest structural weight balanced against the performance desired. Wings of jet aircraft are becoming mere stabilizers as speed increases. Jets are literally propelled, not flown on their wings, as are slower airplanes.

Wings fall into two broad general categories: Cantilever or externally braced. Spar construction and arrangement, plan-form and airfoil used, varies accordingly. A cantilever wing usually tapers in both plan-form and depth and generally has the front spar located more centrally, chord-wise, for carrying a greater proportion of the load. The relatively light rear spar supports the ailerons and flaps (if any). The cantilever wing often uses a box spar of compound construction, or a laminated spar. Those in smaller craft may have a plain but sturdy tapered spar both fore and aft. An excellent example of a laminated-box spar construction (both fore and aft spars) in a cantilever wing as used in an amateur-built plane is that for the Cvjetkovic. The spars are of spruce laminations, top and bottom, internal reinforcing blocks at all fitting locations and complete plywood facing on fore and aft sides. The full leading edge, from top to bottom edges of the front spar, is plywood covered. The balance of the wing is fabric covered only. It is a very strong and beautiful wing. The ribs are standard truss type.

Ailerons and flaps of plywood skinned wings are usually covered with fabric only to keep them light for balance. Weight aft of the hinge-line is to be avoided, particularly in the higher speed ranges.

The current production Bellanca aircraft still uses the very sturdy, multiple-laminated, solid spar in their cantilever wing, with the leading edge plywood covered. The following descrip-

A stack of new spars for Bellanca wings. The multiple laminations can be seen at the still rough root ends.

Bellanca wing leading edge/front spar assembly (right) ready for the formed plywood skin (left).

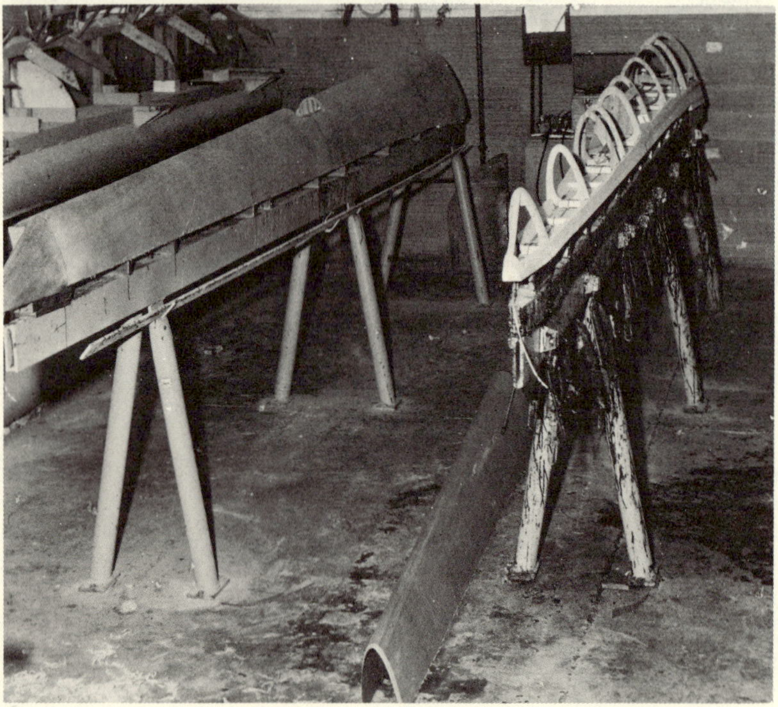

ellanca ribs, spars and top-skin were assembled in a large jig to hold
 in exact alignment. The full skin completes a rugged structure.
ote spruce reinforced plywood compression ribs and the full
ngth top and bottom rib gussets. Formers aft of the rear spar
e secured by corner-blocks. Hole in tip rib for navigation
ht.
ompete Bellanca wing assembly is dipped in pre-
rvatives.
ellanca flap (left) and aileron (right) ready
r cover. Note the minimum weight aft of
e hinge line.

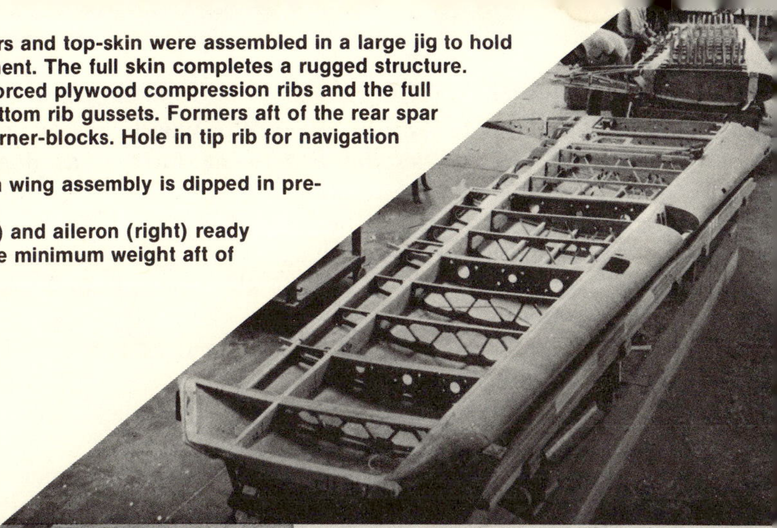

tion is from their Chief Engineer, W. R. Shackelford: "I might point out a few reasons we prefer wood construction.

1. Wood does not experience fatigue, as does metal.
2. When properly protected, it does not age or corrode.
3. It is a natural for bonded construction (bonded joint strength *increases* with age.)
4. It is easily workable with uncomplicated tooling and equipment.
5. Smooth contours are obtained as a result of bonded construction.

There are many other reasons, but these are representative."

"Specifically for the Bellanca construction, our spars are rectangular cross section and are tapered. Basic material is Sitka spruce, with harder woods laminated in at the root for bearing plates. Skin is Mahogany plywood. Ribs are built up truss type made of a mixture of Sitka spruce and Mahogany plywood. Contours, such as the leading edge, are made by steam soaking the plywood sheets and forming them over molds. Bonding is done mostly with Penacolite — a Koppers product. We completely construct the Bellanca wing, then dip the entire structure in a temperature controlled sealer; thus, there are no exposed areas where weathering could occur."

Bellanca Viking wing, all-wood, is smoother, lighter and (said by the manufacturer) stronger than metal wing.

"Our "G" stresses are designed around normal category, 3.8 G's limit 5.7 G's design. Ultimate G's is not known since we haven't carried a test to failure. The airfoil is of Bellanca design. It was developed in a wind tunnel rather than being patterned after any of the conventional NACA designs."

Spars laminated vertically are commonly used and more will be said of them later.

Box spars as compared to multiple laminated or plain solid spars are much lighter and compare favorably in strength/weight ratio. They can be built very lightly for very high efficiency, as in the many high performance soaring and sailplanes. Use of the light metals is now being favored for these, however, and some fine work is being done with plastics. But in sailplanes of wood construction the box spar is the best, providing the necessary strength for the very long span with gradual taper to near zero at the tips.

The justly famous WW I Fokker D-VII biplane and the DR-I triplane had individual cantilever wings (the inter-plane struts were only "moral support") of full-chord, untapered plan-form and section. Both front and rear spars were box-type, with upper and lower spruce beams joined by fore and aft plywood faces. Of course, all "hollow" spars have solid wood at all fitting locations. With the then very advanced fuselage construction of welded steel tube, Warren truss, they were very rugged aircraft. The later Fokker tri-motor and single engine commercial planes also used the deep airfoil and cantilever wing which was one of the biggest steps toward modern construction and advanced performance.

The *box-spar* is used primarily for 1(: a cantilever wing and 2(: a wing with deep airfoil in the inboard portion and tapering toward the tips. A *solid laminated* spar may be similarly used but it is less desireable because of its greater weight. A *compound* type is often an excellent choice although more complicated to build.

Externally braced wings are most common to both monoplanes and biplanes, but especially to the latter. The airfoil will depend on the design and the performance expected. Nearly all biplanes now in use have solid spars, but here again there can be quite a variety:

1(: plain, 2(: laminated vertically, 3(: routed internally or more commonly, externally, 4(: I-beam. The latter two are variations of the same thing, in practical reality. The I-beam section is achieved by external routing as in the D.H.60-G or M and many others of the 1920's, or by adding flanges to both spar faces, top and bottom, to a relatively thin, solid primary beam as in the Fairchild-22 and 24 Series aircraft of the '30's, which have deep spars as compared to the shallow section of the wings of the 1920's. Some of the Travel Air biplanes had internally routed front spars and plain solid rear spars.

Generally, the deeply routed spar was used in wings of thin airfoil section. The spars were relatively wide to accommodate the deep routing (⅜" in the D.H.-60 wings). This makes a lighter and more flexible spar than a solid one of equal external dimensions such as used in the Meyers OTW biplane which also has a similar thin airfoil section. A thin wing is more the "speed" type airfoil; a wing of deep section is "high lift". Of course there are all degrees and variations and to go into the subject leads one very shortly into aeronautical engineering problems, well beyond the qualifications of the author. Our interest here is in the construction.

Vertically laminated spars are rather common. Wings of the Boeing/Stearman PT Series utilized both plain and exterior routed-laminated spars, apparently depending upon which subcontractor fabricated them. The plain solid spars were the most common, but I have worked on many which had two laminations and some which were of three laminations and these latter were made up of short lengths, scarfed and laminated together and having exterior routing approximately ⅛" deep. These spars for the most part were of Douglas Fir rather than spruce. Laminated spars resist checking and cracking through bolt holes, but routing a narrow spar is hardly worth the very slight saving in weight because only a small amount of wood can be removed, and the subsequent fillers needed to mount the ribs replace a good percentage of what has been routed out. So the more nearly a spar approaches a square section, the greater the benefit of routing. For a narrow and deep spar adding flanges, as in the Fairchilds, is worthwhile to arrive at the I-beam spar section.

In any case, gluing and pressing of laminations or of plywood faces of box-spars must be very well done, with even application of glue and firm pressure throughout during the setting and curing time. The Posey Mfg. Co. still has a huge press which they used in years past for laminating spars. Pressure was by screw action and the machine is very heavy and level for producing absolutely straight spars. Any curve pressed in during lamination is there to stay; that is the way we make wing tip bows. In making spars for small wings, wide-flange angle-irons clamped tightly together with large C-clamps on a flat and even table will do quite well. The main thing to watch for is that all edges are even and true. Nails are not used. The writer has made spar blanks in this way for a new little biplane and this was done not only to make strong and lightweight spars, but also for convenience and economy of both time and money. I did not have on hand spruce stock of specified thickness but I did have it in one-half thickness. So, by laminating, spar blanks were made which will be better than the plain ones would have been.

The flight characteristics of an airplane depend then on the design and arrangement of its wings and on the airfoil used and almost every imaginable combination has been used. Anyone interested in design and construction will do well to study the complete history of flying machines. It is wryly amusing to see so many new-comers with an "original" idea that was tried and discarded between twenty-five and fifty years ago, or perhaps developed and currently in use very successfully without his realizing it or recognizing it in its highly developed state. On the other hand, there are most often BETTER WAYS of doing almost everything. For the most part, simpler is better. But the first thing is to understand the problem. The spars bear the primary loads and stresses in flight. The ribs provide the lifting force combined with the forward movement achieved with the power-plant, whatever kind it may be. A certain amount of flexibility of the wings is desirable; it reduces stresses in other parts of the airframe.

A full set of spars for the Hatz biplane are cut and bevelled. Left to right: finished reinforcing plates glued at root end and inter-plane strut fitting area, but root has not yet been sanded clean. Next, inter-plane strut area plated and clamped until fully dried. Rest of spars waiting for plates, rough-cut leading edge strips, wing sub-spars to which aileron spars (next) will be hinged, aileron sub-spars just visible behind the front and rear center-section spars (on edge).

5. Construction of Spars

Spruce is the all-around best wood for spars, being relatively light, flexible and straight grained. Other woods tend to splinter or fracture much more readily, as I know from many years of repairing wings. Small areas of knots or grain irregularity are weak points. In a laminated spar these minor faults (if small) can be somewhat cancelled out, but only choice pieces should be used for spars.

Spar blanks are best ordered to dimensions given, surfaced both faces and edges. They are shipped (by Posey Mfg. Co.) with 1/16″ extra, one edge, to allow for bevelling and finishing as required. The bevelling *can* be hand done, but better on a table-planer or portable router with a planer attachment which can be adjusted for angle of bevel. It is very difficult to get a uniform edge bevel by hand-planing.

At any time when it is necessary to tightly hold spar stock for cutting or planing or drilling, use flat blocks of wood under clamps to protect the spruce from crushing the surface.

Handle all wood carefully to avoid damage to surfaces, edges or corners. Keep all wood *clean* at all times. Use pencil or pen for marking; not any kind of grease-pencil or crayons, or any sharp or jagged tool.

Plywood (or other) reinforcing plates at fitting areas are glued and secured in place with a *minimum* of nails and pressed between wood blocks with large C-clamps or parallel clamps during the setting and curing time of the glue. Reinforcing plates are *not nailed on*. Use only enough nails (No. 20 size of approximately 2x plate thickness) to hold the plates in position for pressing *only*. Pieces of pine about 1″ thick are best to use for the clamping; they must be large enough to cover the plate (each side) and provide an even pressure. The number of clamps will depend on the size of the plates. The object is to get the most even pressure over the entire area. When clamps and blocks are removed, very carefully use a knife or broad wood chisel to clean off glue that had been pressed out along the edges. In *all* aircraft woodwork, clean away all glue on all surfaces as soon as possible, but since the plates must remain clamped on spars for at least twelve hours, preferably twenty-four, the glue will have dried hard before it can be reached, assuming the pressing blocks are larger than the plates.

All spar reinforcing plates (except at the root ends) should have the ends bevelled (feathered) to reduce the possibilities of shearing of the spar where there is an abrupt termination of the reinforcement. Bevelling "tapers out" the stresses. One main objective in aircraft engineering is to evenly distribute loads and stresses instead of allowing them to be localized and concentrated. This bevelling is easily done with a hand-plane, worked vertically, then sanded smooth.

The root ends of all spars must be cleanly finished, best done by holding level on the disk-sander. These ends should be finished with a coat of good wood sealer and then varnished, or at least two (preferably 3 or more) coats for a clear gloss. *Do not paint nor cover with dope or fabric.* They must be open for ready inspection for weather checks or cracks to or from bolt holes. A clean, smooth surface prevents moisture from entering the end grain.

On new spars all holes for fittings should be done after all reinforcing plates are on and bevelled (to match the spar). Holes must be absolutely accurate in size and location. Double-check for precise locations (it is easy to get a spar upside-down!), and use correct drill sizes (neither under nor over). It is essential to use a drill-press with the spar held securely to prevent any shifting and in a level position for exact 90° hole penetration. Be sure the drill goes cleanly through, and use a backing block of wood so there will be clean hole edges both faces. If it is necessary to cut a large hole, as for clearance for a control, use the correct size hole-cutter or fly-cutter. Go part way through from one side and then reverse the spar and complete the hole from the opposite side; the pilot-drill will give the exact location.

Do not use a hand-held drill-motor for drilling spars because it can not be held firmly enough for use with drills over ¼″ without making star-shaped holes, nor can the precise 90° hole angle be accurately held. The only time to do this is in case of repair in the wiring (assembled), where it is impossible to use a drill-press. In this situation, the reinforcing plates are pre-drilled and glued and

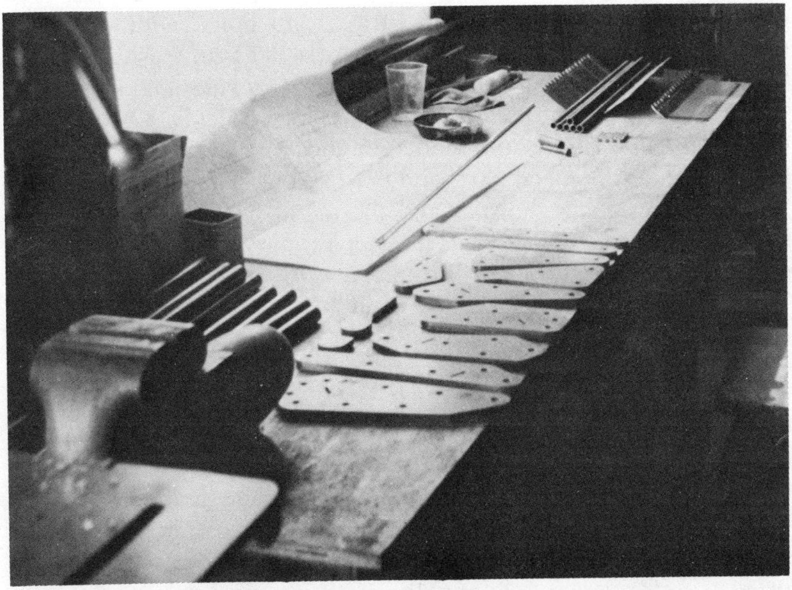

4130 steel wing fittings for the Hatz biplane are cut out and bent but not yet welded. Jigs must be devised for all welding to hold the work and to prevent warping by the heat. All aileron hinges were made later. The plate type fittings in the foreground are stacks of four of each, not clearly shown in the photo. Upper and lower panels require different fittings. All will be sand-blasted and zinc chromated before installation.

pressed in exact location, then drill very carefully through the spar using a clamped-on fitting as guide and template.

Small holes only partially through a spar, such as for small screws, are more safely made with a hand drill; there is less risk of going too far.

When the spar is completed, lightly sand it all over, including corners, to preclude snagging of little splinters which might become big ones. This will give a nice clean surface overall and also make it easier to slip the ribs into place during assembly; it is more satisfactory and a lot easier to sand when there is nothing in the way. The smoother the surface, the better the finish when varnished. A smooth surface holds far less dirt and moisture; it also takes up less varnish.

Varnishing of the root ends of the spars has already been explained. Now, with either new wings or refinishing or repair of old ones, apply at least two coats of varnish over all locations which will be under any metal fittings and bolts (with washers). Do *not* get any varnish where anything (such as rib trusses) will be glued. This is to prevent mildew and rot developing under the metal where moisture can not vent and dry quickly. Overall varnishing *after* assembly does not get the job done. Moisture will work in under the metal parts and not dry out. The wood eventually rots and weakens the spar and the metal will either corrode (if aluminum) or rust (if steel). This is a primary cause of failure of wood structures and their fittings in any part of an airframe. There is no excuse for it. Allow each coat of varnish to dry thoroughly before putting the fittings in place. Brush the varnish evenly and carefully *only* where it is to be under metal so it will not interfere with gluing during further assembly.

While still discussing spars it is interesting to note the variety of arrangements which can be used according to design and planform of different kinds of wings. The average lightplane wings have simple spars with conventional parallel spars in vertical position, but some wings are different in this respect and their proven performance validates their unique spar relationship under aerobatic or other special stresses. The familiar Stearman PT Series wing has its spars parallel in both the horizontal and vertical planes; they are *not* perpendicular to the chord-line, but to the *thrust*-line. Their girder-like compression members very firmly secure them so. The famous aerobatic Waco Taperwing of the 1930's had wings which tapered in both plan-form and section, as did numerous others, but the spar relationship was unusual. The solid spruce, tapered spars of constant thickness, not only followed approximately the outlines of the leading and trailing edges, with wide separation at the root end and approaching each other at the tip, but both fore and aft spars also canted outward from the bottom side. In other words, the front spar leaned forward (from the bottom) and the rear spar leaned aft (from the bottom). With their uniformly tapering ribs and spruce compression members the wood structure alone was very rugged, but of course they also had the standard internal drag and anti-drag wires. The "weak link" of these wings was the arrangement of root-to-fuselage attach bolts which ran vertically through the spars very close to the butt (root) ends. I rebuilt one set of these wings where all four of these bolts had at some time been pulled right out of the root ends. Bent inter-plane strut attach fittings, combined with this damage, indicated a violent snap-roll in which some lucky person had very nearly shed his wings. It is probable that the wings and center-section had not been correctly rigged, but what ever the cause, the flying and landing wires had held the plane together, to be later (much later) rebuilt and flown again. It was an interesting job: one panel built completely new, a couple with one spar replacement and repaired ribs and one with repaired spars and all new ribs. A typical restoration.

There is no need to reject a wood spar simply because of age. If the wood is still sound, without evidence of damage or dry-rot (caused by mildew), a spar can be cleaned up and revarnished again and again. The bottom where moisture may have collected due to insufficient drainage and ventilation is the place to look first for trouble, especially where the bottom rib cap-strips attach. A major cause of wood damage is from over-tightening of bolts at fittings, which crushes the wood. It may be necessary to take up a little on nuts and bolts because of slight wood shrinkage over the years, but this should be just enough to make them snug, not torqued down as they would be in metal. This is true of all bolts in all wood airframe structures with the exception of propeller hub bolts which do have specified torque values.

Comparative rib airfoils
Top: Stearman PT-17, double convex
Center: Typical Clark-Y, straight on bottom aft of front spar
Bottom: Assembly jig for D.H. 60-M rib (same contour as rib) Note wide but shallow spar section, slight under-camber, typical of 1920's.

6. Construction of Ribs

Ribs have been made in an endless variety of ways from simple cut-out plywood to welded steel tube forms. For those of average to large size and section a truss style is most practical and efficient relative to weight and strength. The cut-out plywood rib is comparable to the lighter aluminum machine stamped rib in current production aircraft, while the wood truss compares to the even lighter weight aluminum truss very common to the light plane wings of the 1930's and '40's. The aluminum ribs weigh less and the stamped ribs (with lightening holes) are also stronger and more durable, but in the truss type, a wood rib has the advantage; the aluminum truss ribs are very flimsy. Wood is a little heavier, but it is stronger and certainly more repairable; it does not corrode nor bend. If properly built and maintained and correctly covered, warping is negligible. A cut-out plywood rib is good in a very small or thin wing where the truss type would be rather impractical. A composite style using thin plywood web reinforced with light spruce strips and cap-strips makes a good, strong and durable rib for wings of rather small size. In a large rib, to get sufficient strength to support the thin plywood with spruce strips added, the weight/strength ratio would be reversed and a truss type construction would serve better.

Cut-out plywood ribs of many amateur-built planes are far too heavy. Many people lose sight of the need to keep weight at a minimum in small planes and the performance suffers in consequence. It takes more than brute horsepower for a plane to fly well. A plywood web rib less than ¼" thickness must have cap-strip added to provide adequate surface beneath the fabric covering. This cap-strip would be routed (to *thickness* of the plywood for *width* of the routing) then glued and nailed to the edges of the plywood, top and bottom. Depending on the size and section of the rib, reinforcing spruce strips would be added on one or both sides. An example of this type of rib is again the D.H.-60. The 3/32" plywood web sections (leading edge, between spars and aft to trailing edge) are assembled on a special jig, holding the three parts in their correct positions while the top and bottom cap-strips are glued and nailed in place. Additional plywood reinforcement pieces are added at the leading edge and trailing edge. There are intermediate nose formers between ribs because the ribs are rather widely spaced and there is no leading edge cover of aluminum or plywood. It is a very light wing. The bow is of aluminum tubing and the internal wires are No. 6 diameter. But for a full size, two-place airplane of about 98 hp, performance is quite good and these planes have given reliable service all over the world. Compared to these wings, many wings for homebuilt planes of about 2/3 the size are almost twice as heavy. *Heavier is not better.* It is the quality of construction and materials that counts.

Spar cut-outs in plywood ribs must leave enough material, top and bottom, so the rib will not break here. Spruce reinforcing corner-blocks can be added, each side of the rib, top and bottom, where it crosses the spar edges, as well as the attaching vertical corner-blocks on the spar faces (both spars). All of this adds up to a fairly heavy wing structure. Lightening holes in the plywood ribs must accomodate the diagonal internal wing wires (or other diagonal members) used for torsional bracing.

If a router is used with a pattern for cutting out plywood ribs, the pattern is made about 1/16″ under-size to allow for the collar of the router. This works very well when making ribs of less than 3/16″ thickness as several can be stacked and cut simultaneously (they must be tacked or screwed together at a couple points which will not interfere with the router). A router is necessary for ribs with other than round lightening holes, for uniformity and neatness. A carbide-tungston cutter must be used because a common steel bit is very promptly dulled by the bonding glue in the plywood. For ribs which will have added cap-strip, allow extra plywood web to fit into the routed cap-strip, so the total dimension will be correct. Plywood ribs with round lightening holes can be cut out on a band-saw and have the holes cut with a standard hole-cutter or fly-cutter. Any ribs cut out on the band-saw should have the rough edges finished off to even contour on the disk-sander. Any plywood ribs need additional thickness at leading edge and trailing edge tips for the attachment of the metal by nails, screws or soft rivets, which would otherwise split the narrow rib material.

The plain, ¼″ plywood rib with round lightening holes is no doubt

Cutting out 3/16" plywood web sections for the Gypsy Moth wings with the router. Cutter follows the pattern which is screwed on top of two or three layers of plywood to secure them to the cutting board. For both uniformity and safety all pieces must be tightly held down. Son holds a cut piece. Tiger Moth ribs are of similar pattern but of flat spruce strips instead of plywood web.

the easiest and fastest to make, but it is also the heaviest and most expensive (if aircraft plywood is used). Non-aircraft quality plywood lacks uniformity and fineness for ribs, in my opinion. A composite construction using thinner plywood with spruce cap-strips and reinforcing trusses produces a strong and durable rib which is not too heavy for a small plane; this type was often used in large wings in years past. It requires some work, but it is good.

An accurate jig for building truss type ribs requires thought and careful workmanship. Use a (common) plywood board of ⅝" (min.) thickness, large enough to extend several inches all around the rib. It must be heavy (thick) enough that it will not warp and turn out

Rib jig for the Hatz biplane. Rib stock is ¼″ sq. throughout so jig is made with 3/16″ plywood nailed to the board. Nose form block of rib is ¼″ spruce or plywood.

curved ribs. Sand it reasonably smooth then make carbon tracings (or use an old rib for a pattern if from an old wing) of the rib plan directly on the board. Project the spar center-lines and face lines (of spars) and take very careful measurements of the center-lines to be certain they are true and exact, both top and bottom (spars are normally parallel, but check plans for this detail). Be sure they sit at the correct relation to each other and match the fuselage attach fittings. The distance between spar centers must be precise and should always be checked with the corresponding fuselage attach fittings. A mismatch here is a major problem. The set of front and rear fittings determines the angle of incidence, so it is vital that all fittings and spacing be correct. All of these considerations enter into building of the rib jig. If this is a new plane being built from plans, double check. Mistakes have somtimes been made in drawings on this particular point. Do not confuse center-line measurement with inside/outside fitting measurements.

To make the jig, use small blocks of wood 1/16″ less then the thickness of the rib cap-strip and truss material so the glue will not stick to them when gussets are put on. Nail the jig pieces directly to the board in sufficient number to insure the full and accurate contour of cap-strips and to securely hold each truss piece in its proper position. There is no need for a full enclosure though, in any part. A nicely contoured rib will result from allowing the cap-strip to follow as natural a curve as possible. Begin with the lay-out of the top and bottom cap-strips then fit the trusses in from front to rear.

Gussets on the rib in the jig. (Trusses omitted aft of front spar and both at rear spar; see text.)

According to the rib size and design, it may or may not be necessary to make wood forms in which to steam (or soak) the top cap-strips for pre-bending to make them follow the leading edge contour. This will depend on the degree of bend between the leading edge and the front spar. You will find this out when making up the jig, because a piece of cap-strip will be used in locating the jig blocks where they are required. It will largely depend on the dimensions of the rib material.

If extensive re-use of the jig is anticipated it is adviseable to varnish it to prevent glue from soaking into it at the gusset areas. *Do not wax* a jig for this purpose; it would rub off onto the rib stock and make for very poor glue joints, and also poor varnish penetration later. If the jig is to be used for only one set of wings there is no need to varnish it. Glue which will build up at the gusset points can be scraped off enough between ribs that it will not interfere.

Rib gussets are often carelessly planned and made. They should be laid out to glue and nail to all truss pieces at each joint or cluster with sufficient material for one or possibly two nails per piece with approximately ⅜″ to ⅝″ spacing of nails (maximum spacing is ¾″). There should be no loose protruding corners such as would result when a square or rectangular gusset is used on a T-joint. Gussets may necessarily extend a considerable length along a cap-strip, in which case space the nails ⅝″ to maximum ¾″; if practical, try to stagger the spacing on the reverse side. Too many nails, too close, will decrease the total strength. Some amateur-made ribs have looked like pin-cushions.

Gussets are best cut out in stacks of up to ten layers of 1/16″

plywood. Lay out a batch on the top piece and tack the bundle together with a couple ¾" No. 20 nails. Draw the largest gussets first. Small ones can very often be cut from material left from the corner or inner remains of nose-formers, etc. This is a fun-place to see how little scrap you have to throw away! Just be certain you provide for maneuvering on the band-saw without getting stuck in a blind spot. It is often easy to use a common cutting line advantageously. The same kinds of gussets should have the same grain direction for uniformity. Finish them (still in bundles) on the disk-sander so all edges are clean and even and the dimensions uniform. All edges need to be sharp and clean for sound glue joints. Good glue joints are not possible with rough edges.

Select rib cap-strip material carefully, using the best straight and even grain for the top, then the bottom, full length strips. The top piece will have the most curve over the leading edge. Defects could cause breakage over curves of substantial bend. Little defects can be eliminated when cutting the short truss pieces; but if minor, they can be disregarded. Certified Spruce seldom shows any defect to be concerned about, but pick the choice pieces for the full length strips. These usually have the leading and trailing edge ends cut off at an angle to fit the nose block at front and the opposite (usually bottom) cap-strip at the trailing tip. I add a wedge-shaped block between them at the tip for reinforcing and to prevent warping.

Cut and fit *all* cap-strip and truss pieces so the grain is *vertical, not flat.* Because, regardless of the kind of covering material used, if the rib grain lies horizontally instead of vertically, the cap-strip will warp and flatten between trusses, resulting in the loss of the original airfoil contour, and therefore, the lifting capability of the wing and loss of performance of the airplane. With rib grain vertical the wood will resist this flattening tendency and hold its proper airfoil contour. Properly covered, the ribs will not become warped. (See: *Aircraft Dope & Fabric* by Spencer) In addition to the above reason, the nails hold much better going through (or across) the grain than when driven parallel (or between) the layers. An exaggerated example to illustrate this roughly: push a nail *through* corrugations of cardboard, then *between* corrugations. Nails between layers would at the same time tend to split the wood. With all rib material set with the grain vertically, the surfaced edges of the wood will all be in the same plane and the width will be absolutely uniform throughout.

Lay out and cut in advance all the parts for the full set of ribs and arrange them for convenient access during rib assembly. This will save a lot of time and confusion. Make separate (identified) piles for each of the truss sticks on a large board near the work-bench and keep the various gussets in neat piles or separate little boxes handy on the bench. This saves a great deal of picking and sorting. I assign a sequence number for every truss member, from leading edge back to the rear tip. There will be a separate group of both truss sticks and gussets for the aileron ribs, and this brings us to another note in making the rib jig.

It is obvious that the aileron rib is exactly like the primary ribs except for the aft portion, so with a little extra planning the one jig can be used for both by simply laying out the aileron rib truss locations when making the jig. Sometimes there is interference but often both can be provided for at once. Additional locator blocks can be nailed into their respective places, with the wing sub-spar (to which ailerons generally are hinged) and the one or more aileron spars allowed for as were the main wing spars, and with the same care. It will probably be necessary to use rather small pieces for the jig forms so there will be room for both lay-outs. Trace the rib patterns on the board for both types of ribs to visualize what will be needed. If it does not work out feasibly to make the dual-purpose jig, make up the full, regular ribs first, make the minimum jig changes and assemble all of the aileron ribs. They will be in one piece as they come out of the jig. The drawing will have taken into account the hinge space.

Leave the aileron ribs intact until after sanding all around the periphery (all ribs) on the disk-sander for a clean and even contour. There will be small edges and corners of gussets and glue squeezed out and light sanding of the rib all around will remove all of this. Just be careful not to make flat places or hollows which would spoil them. After this contour sanding, cut off the aileron portion of those ribs. The ailerons will of course be assembled separately from the wings.

But let us return to the basic rib assembly. If every detail of construction were perfect, every rib would fit both spars perfectly with exact final spar-center to spar-center dimension with the internal rigging completed, but it seldom works out quite that well. Nor could the completed ribs be slipped on past the spar reinforcing plates which most wings have, usually in several locations. This is

Modification of jig for aileron ribs. Ribs will be left intact until ready for installation. The full plywood covered aft sections go at the inboard end of the aileron and the aileron well of the wing is similarly covered. The parallel pencil marks indicate location of the aileron sub-spar. Note the simple primary tools that are essential for aircraft woodwork.

an obvious complication; the other problem becomes obvious only after trammelling of the wing. When the internal drag and anti-drag wires are tightened, ribs that may be just a bit (thousandths of an inch) long between spars are buckled. The only cure is to relieve the compression by removing and then replacing the vertical trusses secured to the rear spar faces. To avoid both of these troubles, *omit* the vertical truss for the *rear* face of the front spar and *both* of those for the rear spar. Just set aside these truss pieces and the gussets until later on and put away the rib jig. This will also obviate the reverse trouble of getting the spar center-to-center space a bit short, as you can see.

There is no way to make any such adjustment when assembling cut-out type ribs to spars so the spar openings must be accurately located and of correct dimensions, but just a *little* over-size to avoid binding and buckling. Corner-blocks will be used in the four corners at both spars to secure them so a slight gap at spar openings would be negligible. It is most important that the measurement between spars is exact at all times. These ribs will have to be put on the spars *before* the reinforcing plates unless the arrangement is such that this is not a problem; some spars have them (plates) only at the root end.

Use nails of proper size in *all* wood construction. For ribs, nails which are as long as the cap-strip is wide are OK, but for any cap-strip in excess of 5/16″, there is no point in using nails over that length. Using ¼″ or 5/16″ cap-strip and truss material, the 1/16″ gusset thickness will leave that 1/16″ in reserve, not penetrated by the nails. No nails must ever go through the base wood any place. In ribs, remember there are nails going into both sides in just the same places. In case of cap-strip of wider dimensions, such as ⅜″, the 5/16″ nails are adequate for holding the gussets; after all, ¼″ nails hold just as well on ¼″ cap-strip.

Aircraft nails are coated for rust prevention and they are finely barbed so they will resist backing out. Except for occasional uses, such as for nailing in rather large corner-blocks, etc., there is no need for nails larger than No. 20, and the larger size, No. 18; these would be nails of 1″ length or more.

One of the most common faults in amateur aircraft woodwork is the use of too many and too-large nails. There are two purposes for nails: to hold two or more pieces in the correct relative position while pressure is applied with clamps and to provide that pressure when other means of doing so are impractical. In the first instance, for example the spar reinforcing plates, the minimum number of nails are used; clamped on blocks of wood do the pressing. Sometimes, usually during repairs, small C-clamps and strips of wood (thickness of the cap-strip or greater) may be used in pressing glued scarf-splices in ribs. The one time when they are used in large number and without regard for spacing is in the gluing and nailing down of "boxed" compound-curve plywood portions at the forward quarter of the wing tips, etc., where the plywood is wetted on the outside surface, then glued and nailed down securely over a comparatively wide area and the only way to make it lay down completely is by the liberal use of nails (½″ max.). It will be nailed along the straight spar edge first (if at a curved bow), then down over the rib(s) and over the curvature of the wood bow. (See Finishing the Wing Tip Bow)

Too-large nails can split the wood, and of course this is relative; little No. 20's can split ¼″ cap-strip if they are set too close to the end of a piece. Replacing small nails with larger ones is *not* the way to secure loose ribs in an old wing. This only splits the cap-strip *and* the spar. Carefully remove the old nails, clean away all old glue and debris, then re-glue and re-nail (avoid old holes) with no larger than No. 20 gauge nails not exceeding ¾″ long.

The only larger diameter nails required are for securing the wing tip bow at the spar tips where a standard No. 6 finishing nail is OK, centered at the outer radius of the bow where it will not interfere with the contouring of the bow (with a hand-plane) after it is in place.

Aircraft nails are flat-headed, barbed and cement-coated, as stated before. Brass nails are seldom used; they are bent easily because they are quite soft and they are of larger diameter, which is not good. They were much used in the past because rust-proof steel nails were not readily available. I can see no advantage in staples instead of nails but several reasons to prefer nails: they are just as easy to locate accurately with the magnetized end of a tack-hammer and the small head of the hammer definitely gets into tight quarters better; staples do rust (unless stainless steel) and they are very poor when used directly under fabric cover; they show and of course the fabric will wear where it rests on them. With a little practice it soon becomes easy to start the little nails with the magnetized end of the hammer and then drive the nails with just a couple firm, light strikes with the driving head. The work *must* be solidly backed up. When working on ribs in a wing, or anyplace other than on a solid bench, use a convenient sized steel bucking bar in one hand and the hammer in the other. If the work is not firmly backed it is not possible to set a nail solidly and if it is not, it is useless. The No. 20 nails will bend all too easily but are far less likely to do so when firmly backed up. Remove any bent nails and replace. Do not mash them into the wood with excessive hammering. A couple of sharp firm strikes are far better than timid hammering; the nail is driven completely and the wood is bonded well. The flat heads of the nails should be just flush with the wood surface. Be careful not to mash the wood surface with excessive hammering. The nails are driven very lightly.

Rib gussets are sometimes poorly planned and can often be improved upon by the builder. In restoration of old airplanes, it is best to duplicate the original size and grain direction, but many amateur designers seem to think that "bigger is better" and this is not so. Do not mistake this to advocate tiny little gussets which would not cover enough area to provide support; my concern is rather with incorrectly shaped gussets with loose corners protruding which contribute nothing. Very often, a triangular or trapezoid shape better fits the truss joint or cluster, with every part (corner) glued and nailed to a stick and securing them all. They will vary from styles of triangles, rectangles and trapezoids to fit the particular groupings of

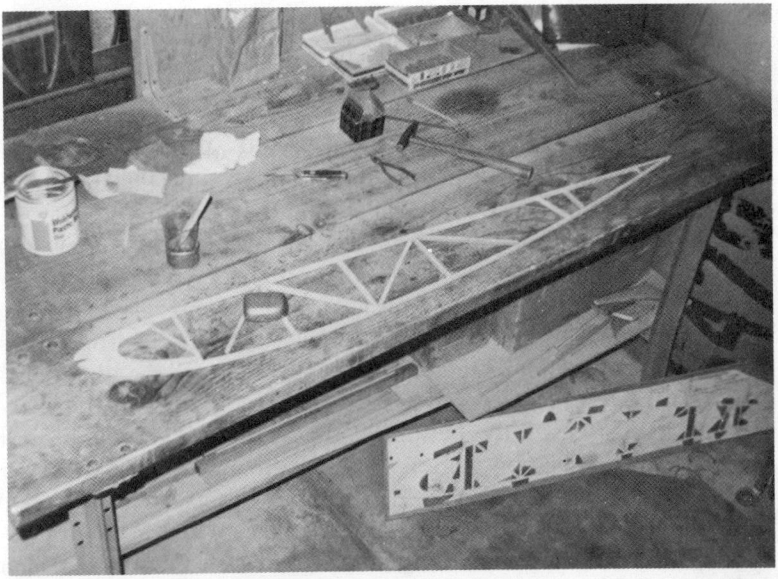

Aileron rib removed from jig for gussets to be put on the other side. Iron block keeps rib flat on the bench until all gussets are on; this is my all-purpose bucking-bar:

cap-strip/truss connections. With a little care in laying them out before cutting, there will be a minimum of plywood wasted. Note the shapes used in the ribs illustrated.

Assemble the ribs by putting all gussets on the first side (omitting vertical trusses at spar openings as previously mentioned), then very carefully remove the rib from the jig, lifting out the trailing edge tip first, working toward the leading edge. It may be necessary to pry lightly here and there with the tip of a knife blade. Work it free as you lift it out, keeping both top and bottom cap-strips coming as you go. Remove the jig from the bench and lay the rib flat on the reverse side. If there is any tendency for it to lay at all curved up, put a bucking bar or similar iron block on it to keep it flat while putting the gussets on the second side. It is often necessary to hand-hold the cap-strips together over the curve of the leading edge until the nose gussets have secured them, because they will tend to straighten out until gussets are on both sides. The spar gap (opening) must be maintained uniformly. If all gussets have been cut and put on with care on both sides the ribs will all be true and uniform. Either hang or evenly stack the ribs to dry. It takes about an hour to completely assemble a rib (less spar truss pieces).

There is much less likelihood of damaging a rib by immediate removal from the jig after putting the gussets on the first side instead of allowing it to set over-night when the glue that has been squeezed out at the gussets would dry and stick to the jig. So the home-builder's common fallacy of "a rib a day" is definitely *not* the best idea. This has been learned through personal experience when an occasional partially finished rib has been left in the jig because a batch of glue was used up about the time to stop working. The rib comes out of the jig much better *before* the glue dries.

When all have been assembled, sand the contours on the disk-sander and also check the cut-out at the leading edge which most ribs have for the leading edge nailing strip to fit in. This will vary in size but it is usually about ⅝″ x ¾″, with the greater dimension horizontally, with about ⅛″ protruding in front which will be planed after assembly to the leading edge contour so the metal which nails to it will have a smoothly rounded radius. If this is set completely into the notch of the ribs, without a portion left for radiusing, it produces a squared-off and blunted leading edge which is *not* according to design of the airfoil.

During *all gluing* at any time, *wipe away excess glue* as you work. There should be NO messy glue left anyplace. This seems to be difficult with the resorcinal types of glue, but it can be scraped clean if it is done promptly; we have seen those which have and which have not been. The Weldwood does remove (still wet) much more easily and cleanly. If excess glue remains it looks very sloppy and if it later flakes off (as exposed glue often will do) it leaves bare, unvarnished wood subject to deterioration from moisture and dirt. So sloppy work is poor work.

The writer has used Weldwood Plastic Resin, water-mixed glue exclusively over many years of wood wing building and repair and I can see no advantages in the various other kinds now on the market. Some of them are very good bonding adhesives but have very undesireable features for aircraft woodwork, such as required catalysts and very critical mixing proportions and conditions. With Weldwood, a working temperature of 72°F (min.), water for mixing with the powdery glue in a small cup (paper or plastic, which can be endlessly re-used) is all that is required. Working time (up to three hours), temperature and mixing proportions are all very much less critical. A neater job results because it is so much easier to wipe off excess as you work (using paper towels). It holds very well indeed, and it lasts the time of the airframe. I have never had a glued joint fail, and I must admit to having tested some the hard way, several times, with my own home-built plane! Read the instructions on the can; mixing is very simple. Add only enough water to make a heavy cream consistency; stir until completely mixed. Keep the can of powdery Weldwood tightly closed so it will not absorb atmospheric moisture. If this should happen it will become caked and should not be used. Mix only a small amount at a time; just what you anticipate using immediately (learned by experience). It has a working life of up to three hours; do not use after it becomes thick and heavy. *Never add water* to a thickening glue mix. Throw it out and mix a fresh batch. It is generally used in such small amounts that the waste is negligible. A small brush serves well for applying glue to an extensive surface such as in making laminations or putting on plywood skin, but for ordinary use, a piece of wood, tapered flat, works very well and it can be repeatedly sanded down for even spreading of the glue. A brush must be rinsed absolutely clean after each use so old glue will not spoil a new mixture. Mix the glue in a cup that is either really clean or one in which a previous batch has completely dried (and does not come loose from the cup). There must be nothing lumpy nor flaky in the glue which could or would prevent a tightly bonded joint.

Paper towels make the best wipes; they are expendable and cheap. Glue sets up hard on rags and makes them useless when working with glue or varnish.

A good glue line in aircraft work should be but a faintly discernable hair-line. The two pieces of joined wood are to be tightly *bonded,* not just "cemented" together. This requires an even pressure, by nails in gussets or by clamped blocks on spar reinforcing plates, spar spices, etc.

Do not remove nails from gussets or other joints "to keep down weight", which would amount to less than two lbs. in one small wood airplane. It is true that their primary purpose is to provide pressure while glue sets, but it is also true that they DO help hold things together. I have repaired old wings (from the days of Casein glue) where *only* the nails were holding. Weight saved is simply negligible. Far more damage is done to the wood by pulling out the nails than any imaginable advantage of removing them. The one time they are removed is when they come off along with the nailing strips used in putting on plywood skin. Where a plywood "box" is put on, as at the forward quarter of the wing tips and for nailing on a plywood leading edge, nails are used in considerable numbers to secure it. They are rasped down along with the plywood edges, to contour, and the surface is sanded smooth.

Whenever it is necessary to remove nails, especially in repair work, use very small, fine, diagonal cutters which will get in under the nail heads with minimum gouging of the wood surface, and lift out the nails—*pry* would be a more accurate word. These little "dikes", along with a good pocket-knife (for countless jobs) and the small tack-hammer and small bucking-bar, are the indispensible hand tools for wing repair, or any other aircraft woodwork.

Ribs are *not varnished* before wing assembly.

We have been assuming the materials were ready for building, but now we had better go back to preparation. Cap-strip spruce is available in good sized bundles (from Posey Mfg. Co.) surfaced two sides to thickness required in flat boards of random widths and lengths (specify minimums) and then ripped to the width needed. For example, you buy ¼″ flat stock and then have it ripped to 5/16″ width, if plans call for ribs ¼″ thick by 5/16″ wide. Be certain to distinguish between thickness (or depth) and rib width, before ordering cap-strip stock. Do not have all the flat stock ripped but save some for the many additional uses that turn up in building any airplane. Many have solid root-end ribs of solid spruce and there will be other items, such as nose-blocks for the ribs, etc. You can purchase ⅛″ spruce for laminating the wing tip bows. If you or a friend have the use of a good table-saw, most of this cutting and ripping can be done yourself; otherwise, it is best hired done at a good mill or cabinet shop with a minimum of mistakes or waste. When material must be ordered from a long distance, "goofs" become expensive in both time and money. There is a lot less waste in buying spruce in specified thickness than by getting thicker rough stock (cheaper maybe) and cutting it to your needs. The average saw takes a bite varying from 1/16″ to ⅛″. Buying in specified thicknesses, you use what you pay for instead of paying for wasted wood.

Use a disk-sander to finish the cut edges of ripped cap-strip and other pieces before cutting to required lengths. This will provide smooth, even surfaces for gluing. Otherwise, there cannot be good bonding surfaces and poor glue joints would result. If rib stock is cut just *slightly* over-width, sanding will bring it to correct dimensions. In regard to sanding, *Manual 18* specifically said *not* to sand before gluing joints. I can see no justifiable reasoning for this, unless they meant hand-sanding where it is very easy to get hollows in one or both surfaces to be bonded, which would make a bad joint. *Machine* sanding will produce a far more even surface than any saw-cut edge and the fuzzies of such an edge to *not* make a good glue joint. Anything at all which is in the way of two even surfaces will make a poor joint. *Do* remove all saw-dust from sanding surfaces to be glued. You are not making a "fill", but a bonded joint.

There are some special problems in making ribs for tapered wings and for wings with sweep-back. I made one jig-board do for all ribs of a Waco Taperwing by beginning with the smallest (tip) ribs, then progressively enlarging the jig on the board with the lay-out for each larger rib, using only enough blocks nailed to the jig-board to maintain correct contour. Since there were only four of each size to be made there was not too much concern for the durability of each set-up. When three or four dozen ribs must be made, all alike, the jig must be well made to maintain the uniformity.

Ribs for wings with non-parallel spars or with sweep-back must have the trusses which attach to the spar faces cut at the appropriate angle to fit the spars with the ribs set in line with the thrust-line of the plane, in other words, in alignment with the slip-stream. This applies also to ribs in the tail surfaces which may have other than straight-through spars. In these wings, it is obvious that it would be much easier to build the ribs in the jig omitting the spar attach truss pieces, as described previously, then putting them in with their gussets after trammelling the wing. They must be bevelled accurately so the angle of sweep-back matches the rib line without causing any bending of the rib with everything tightly in place. Of course, the same is true for the fit of all corner-blocks used in any wing at any place.

This bevelling is done best on the disk-sander with an adjustable tilt-top table which can be set as desired. Work on the side of downward rotation so the work is not snatched up. The work could be damaged and so could you. Work done on *any* machine must

Example of aluminum stamped ribs and spruce spars; for the Great Lakes 2T-1A Sport/Trainer of 1932. Pictured are lower wings.

be very firmly held.

So there is a little more work involved in building other than straight wings. Every glue joint must be carefully fitted and it is more difficult with these angles to be matched.

Assembly of the ribs on the spars will be taken up in a following chapter.

7. Making Bows

Simple, one-piece tip bows of ash were made for the older light-planes by steaming and bending. Most of the old biplanes had them made in this way and they were fine. The ash had both strength and resiliency and it is to be highly recommended. The problem of using it lies in the need for steam bending. Laminating thin strips is the easy way out, and that is what we have been doing.

Aeroengineers tell us that tip bows are obsolete and inefficient and they may be right, but a lot depends upon the kinds of aircraft under discussion and the type of flying to be done, all of which belongs in the design stage. Borrowing ideas from new, advanced styles for application to old is not always successful, nor is the reverse any more so. Bear in mind that the really good airplane, old or new, has been designed as a unit for certain desired performance within a specified range of cost. If you have too many items you wish to change in a given set of aircraft plans, try your own original design. By the time all construction details are worked out, you may wish you had stayed with the first one. The element of practicality plays a large part in the development of building any plane. Some ideas sound great but when the chips are down, they really are not worth the work, at least for this particular project. This attitude may be conservative but it comes of years of watching over-enthusiastic amateurs botching up otherwise nice airplanes. It has been particularly evident in aerial agricultural work. A few really good ideas have been worked out and generally adopted, but many were not worth the time and money put into them.

It began with standard round, tapered tips, and has gone through everything that could be dreamed up on every type of plane men could build a hopper into: plain cut-off ends (shortened or extended), "spill plates" of simple to weird styles, and in recent years, the deeply contoured "lifting" tips. All of these may well be helpful and worthwhile for *specific purposes,* but before arbitrarily deciding to use them, understand their value, that is, why they are used for certain aircraft. What may be good for one type of airplane may be very undesireable on another. Remember also that advertisements are for the purpose of *selling* and very often the ad is overly optomistic in its statements.

One thing is in mind in this discussion: do not be persuaded to trade wing span for any claim for fancy tips. A good basic wing with a conservative span/chord ratio is the best safety feature you can build into your plane. Remember: an airplane flies on its wings. The engine makes it go, but if it quits, or if you get wrapped up in an overly tight maneuver, you will fly out of it with your wings; and an unstable style of wing just won't be very much help. If it is aerobatics you are interested in, the standard rounded tip is still hard to beat.

Tip bows and biplane center-section trailing edge bows are laminated in jigs. (See photos) This sounds simple but the work is in making the jigs. The full contour, both inside and outside, is laid out on a fairly heavy plywood board large enough to accommodate both the inner and outer forms which do need to be continuous, rather than blocks as with the rib jig, because considerable pressure is required to form and press and bond the laminations. There can be up to ten strips of ⅛″ wood to be formed simultaneously and this takes *pressure*. Any gaps between form blocks would permit flattened areas in the bow contour.

The form-blocks (pairs of inner and outer, mated) must be deep, or thick enough to cover fully the greatest depth of the bow. For instance, that for the Stearman is over 2″ deep at the front end of the tip bow and 3″ or more through the central part of the center-section bow, so the forms must be in excess of these deepest portions.

Bolt the inner form blocks to the board, allow space for the full width of the bow plus the outer set of blocks with an inch or so margin, and cut away excess board so there will be room to turn the handles of the large C-clamps used for pressing. The form blocks will be cut from 2″ x 4″ or larger, as many as it takes to completely outline the bow on each side. It may be necessary to cut notches in the blocks where a wide radius would not let the C-clamp reach across the pair of blocks with the bow material in between. Wider reach parallel clamps help here, but they are expensive to buy. Those in the accompanying photos were made of 2″ x 6″ blocks cut in sections to make up the required contours.

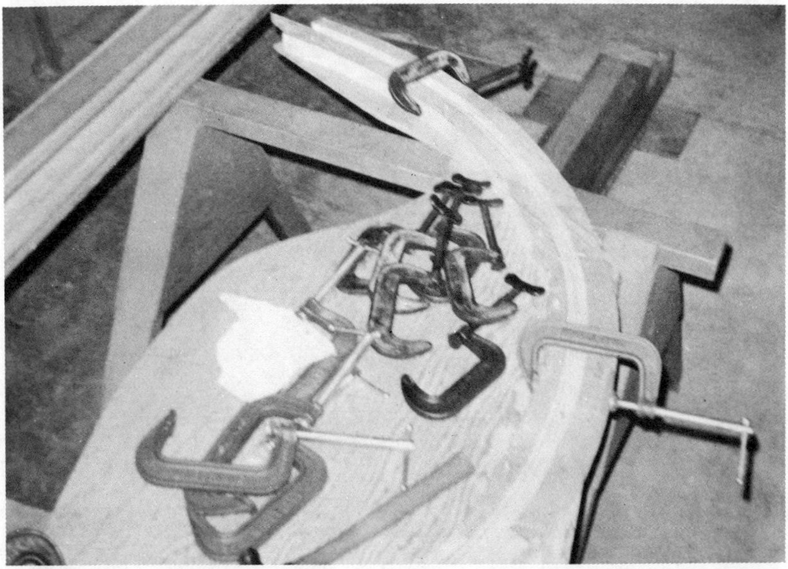

Inner form blocks for wing bow jig bolted to board, outer blocks tentatively clamped while the glued bow laminations are set in place. Hammer and block of wood tap them down evenly onto the board.

These were simple bows of only 1″ x 1″ finish dimensions. They will be planed to rounded shape *after* being secured to the wing. (See chapter 8, "Wing Assembly".) A large bow like that of the Stearman biplane, has added laminations staggered into the inside radius for a total of about ten at the trailing edge portion where it tapers down quite thin and flat. Without these added laminations in this aft part, it would be very thin and flimsy. Because of the taper from over 2″ deep at front to only ½″ at the rear, much excess can be cut away on the band-saw. This must be done with consideration for right and left and top and bottom sides. Do not try to saw them to anywhere near exact fit; leave the final shaping to be done on the wing with the bow secured in place. The top and bottom contours do differ; the bottom is flatter, matching the rib contours. Cutting away the obvious excess makes the bow easier to handle.

The ⅛″ spruce strips for the bow laminations bend easily enough that it is not necessary to steam or wet them before they are pressed into the forms.

Clamps are put on and tightened from leading edge (far left) end with clamp spacing as even as is practical. A deeper bow (at front) would require thicker form blocks. Excess at ends will be cut off. Jig should extend an inch or so beyond measured length to ascertain adequate bow length.

Center-Section trailing edge bow set up in its jig. Wing bow jig at the left.

Shown, from left: complete wing bow jig, center-section bow jig and the center-section bow as removed from the jig. Shaping and planing is done on the assemblies for proper contours. Hardened excess glue is removed with a coarse file; it would promptly remove the cutting edge from a plane.

Putting glue on all of these strips calls for fast work because they must all be done and the clamps tightned on the form blocks before the glue starts to dry. Mix plenty of glue before starting so time will not be lost mixing an additional amount, and have everything ready for the whole procedure: blocks laid by their respective locations, clamps close at hand, opened up fully.

As glue is put on each successive strip laid one on the other, keep them stacked evenly and as flat as possible. This is one time to put on a generous amount of glue; the cut surfaces of the strips will absorb it quickly. Carry the stack to the jig and secure the leading edge of the tip bow first and work aft. Do not tighten the clamps until all are in place. Use a block of wood to push the laminations down so their upper (visible) edges are even; be certain that the clamps are set well down as close as possible to the board so pressure on the blocks is centered (so edges of the blocks are in parallel). Do not be concerned with all the glue that is squeezed out because all the roughness will be completely cut away later anyhow. With the center-section bow, first secure the *center* portion and work outward with the blocks and clamps.

It is surprising how much length of the strips is taken up going around the outer radius. Each succeeding outside piece is inches shorter as it approaches the trailing edge, so in laying out the material for them, first measure the length around with a tape rule and then try a cut piece of spruce around the outer radius — and cut the strips with *extra* allowed.

After all clamps are in place and tightened enough that everything is even, start at the front again for the tip bow and the center for the center-section bow and tighten all the clamps by hand as evenly as you can. There is no need for excessive pressure; you do not want to crush the wood.

Allow the bows to set up in the jig for 24 hours for thorough setting; longer if convenient, especially for larger size bows. It may take some persuasion to get the bow loose from the jig because of glue squeezed out the bottom edges onto the jig board. One way to avoid this is to put a layer of wax-paper down on the board at the time the form blocks are put on. However, this would come up with the first bow out of the jig. Better, wax the board before each one is set up. Since all of the edge will be cut away, no harm will be done by the use of wax here. It is of use only in keeping the jig clean, actually, as the bow itself will not be damaged with the slight amount of prying needed to get it loose from the board.

Final contouring and finishing of the bows will be described in conjunction with the assembly, and will include making the hand-holds. Another thought in regard to tip bows, however, is provision for wing tip navigation lights. Either an enlarged area can be made in the laminated strips at the light location, which would require deepening the jig also, or by adding a shaped block to outer radius of the bow before the final shaping. A hole through the bow will be required for the part of the light fixture which extends within the wing and carries the wire connector. There has to be sufficient material left around this hole to maintain the strength of the bow. The base will fit the shaped flat area on the bow and be attached by screws. The alternative is to use an old-fashioned style of light fixture which clamps around a tube-type bow, but these are not as nicely stream-lined. This is something to consider and plan for before building the bows, and they must be properly located to meet given specifications for angles of visibility.

8. Wing Assembly

With the new spars all ready for assembly, there is one more thing to be done before slipping the ribs on to their individual locations. According to the plans, or as measured from the original wing if this is a rebuild with an old one as pattern, mark the exact place for each rib on the bottom edges of both spars for each side of the cap-strip; a center-line mark would be covered by the rib so its position could vary either side of center. Using a square, extend the lines up on both spar faces for true perpendicular references when gluing and nailing the ribs when the time comes. The same lines can be extended over the top edges of the spars if desired but it is not necessary; the guide lines on the spar faces can be seen from the top. The bottom edge marks are the most important because the wing is assembled with the bottom side up. It will not be turned over until after trammelling if this assembly procedure is used.

Assembly is done with the wing upside-down because greater accuracy of measurements is possible in this position. The spars have less (or sometimes no) bevel on the bottom edges and the lower edges are generally in more nearly the same plane on the lower side of the airfoil. At the same time the rib locations are marked, also extend across the lower spar edge the accurate center-line of the compression member attach fittings and precisely measure and mark the *spar center* of each — both front and rear spars for every compression member. These are very important to the internal rigging of the wing panel. It is unnecessary to mark the top edge as the rigging (trammelling) procedure will be done from the lower side. Precise measurement of the distance between spar centers and between the compression members *must be accurate* throughout the length of every wing panel, for mating with the fuselage and for the wing to be square and true within itself.

The compression members must allow (in length) for spar reinforcing plates and other under-lying fittings so there may be a variation of lengths of the compression members for each wing and care must be taken to put them in the correct locations. In some wings they are interchangeable. The lateral distance between them may vary considerably, so the lengths of the drag and anti-drag wires would differ accordingly. The root compression member of straight wings will be set in place with any plate-type fittings at that location and this will hold the pair of spars in upright position. The other members will not be set in until *after* all of the ribs (except the tip one) are hung loosely in place. Be sure to have each rib in proper sequence, for instance in a lower wing panel, the reinforced ribs of the wing-walk section will go on first, then the few regular full length ribs between wing-walk and aileron, then the aileron ribs. The tip rib is usually outboard from the last compression member and it can be set in place after trammelling, before the bow is put on. Ribs of an upper panel will go on in the same order; here there may be shorter ones if there are flaps (ribs would be made up as per the aileron ribs), then the aileron ribs for a monoplane. The upper wing for a biplane will have ribs all alike except for the tip ribs. A biplane lower panel may have several special ribs, so be certain each is in the correct place as they go on, and remember: the wing is bottom side up. Without the spar-face trusses in the ribs, they can be maneuvered past spar reinforcing plates, and with only the root-end compression member secured, the spars can be moved fore and aft which helps in sliding on the ribs.

Other than straight wings with constant chord and spar depth and spars parallel are a special problem and assembly amounts to putting the spars *into* the ribs. Pull all of the ribs onto the front spar first and support this assembly in a position with the spar vertical. Then start the tip end of the rear spar through the opening of the root rib and work it on out very carefully through the balance of the ribs. The root fittings and compression member will then be secured before proceding with installation of the rest of them.

With the ribs loosely in place, all metal fittings associated with the compression members are bolted into their respective places. Minor amounts of less than correct compression member length can be shimmed with wide-area washers of 1/32″ or 1/16″ to maintain the specified spar-center-line distance, but in no circumstances *over* this measurement. There will be differing compres-

William Slater's CA-65 stabilizer and elevator assembly, with elevator trim tab (right) has plywood ribs with spruce cap-strips. The openings in central skinned areas are for access to hinge bolts and nuts.

sion member lengths to accomodate spar reinforcing plates in most wings. Measure the spar center to spar center distance as each set is tightened and shim as necessary. This is essential for a trued and square wing — which every one must be.

Other fittings not essential to the basic structure need not be put in until later on, after the ribs have been secured; they would only be in the way of gluing and nailing the rib trusses and gussets.

Some wings have wood reinforced compression ribs or box members and of course the same accuracy of spar-to-spar distance applies.

Full-skinned wings may or may not have internal torsional wires or other diagonal bracing as the skin is intended to carry these stresses. If there is not any internal bracing, the wing structure must be fully jigged very accurately before gluing and nailing all ribs in place permanently and the skin put on the first side. All reinforcing corner-blocks would be put in along with diagonal bracing at root, tip, etc., and all control mechanisms, tubing (for air-speed or hydraulic lines), wiring, lights or other systems built into the wing. The inside of the skinned wing will be completely varnished, at least two coats, with the same being brushed onto the inside of the top skin pieces *except* where it will be glued to cap-strips and spar-edge nailing strips. Brush the varnish care-

Fin for CA-65 with skin on the first side. Interior and inside of second cover piece is fully varnished except where it will be glued and nailed. Internal diagonal bracing is unnecessary with the full skin. Note semi-box type spars of fin and rudder.

fully into every corner and pocket; brush it out and away from low places to which it will run. Do not let it accumulate in thick globs anywhere. Over-lapping scarfed edges of the plywood skin sections must be carefully mated for perfect fit. Skinning the wing will require at least two people for spreading the glue on all parts as quickly as possible, then securing it with nailing strips over all adjoining pieces. Note here that nailing strips (same thickness as the cap-strips) must be put on the top and bottom edges of the rear spar as well as the front spar on skinned wings, and they will also have a wood trailing edge for the same purpose. The skin will have to be held down by weights or some other means until nailing is completed so the glue will not set before this can be finished. The job calls for team-work.

Look at the photos of the very neat, clean work done with the tail surfaces (done as above) for the home-built Cvjetkovic CA-65 by William Slater. The tail surfaces illustrate nicely the way this type of construction should be done. His metal fittings match the woodwork throughout.

Return now to the more conventional wing which uses some type of internal wires with adjustable length attach fittings of which many kinds have been made: wire and turn-buckles, threaded barrels like the J-3 Cubs, right and left hand threaded fork-ends or wires going through the spars having wide-area washers and nuts tightened against hard-wood blocks on the outer faces of both spars. Holes for these must be carefully drilled for correct angles for clearance of the wires when they are tightened. Here, as everyplace else, wide-area washers must be used under nuts to prevent any crushing of wood.

Whatever type you are using will be installed now, with ribs loose and compression members secured in position. It is adviseable to put all right-hand threaded ends either fore or aft instead of mixing them, for consistency and ease of adjustment.

Ready for fabric cover, the fine workmanship by Mr. Slater is apparent in the rudder structure.

Fittings are bolted in place and the drag and anti-drag wires are installed and ready to be tightened. The trammell bar and set of points are laying on the wing; the near set of points has fine adjustment.

It is aggravating to have them mixed when you are trying to make minute adjustments of maybe 1/32″ (close enough) and find yourself turning the thing the wrong way, either loosening or tightening when it should go the opposite way. I just make a habit of installing the right hand threaded ends at the front spar and the left hand ones at the rear. It saves confusion and time.

At first, tighten all of them only enough to take up excessive slack. The exact center-line points of the spars coinciding with the extended center-lines of the compression members are the reference points for trammelling. Mark them with point of pencil or a scribe. The latter is better for setting the trammell points while reaching across the wing.

A set of trammell points mounted on a straight metal bar 4′ to 6′ long is best for precise measuring but a measuring tape can be used. You are not concerned with any specific distance but in getting the *same* length per pair within each bay (between compression members). If the hypotenuse of the triangles formed by the wires within any square or rectangle equals the mate to it, with other sides equal, the whole will be square (in line). If the compression members are not the same distance apart the diagonal measurements will differ from one bay to the next but this is of no importance.

The comparative measurements are taken diagonally across each bay of drag and anti-drag wires, which provide the torsional strength as well as serving to square the wing. They *can* be tightened too much and this will pull the wing out of shape and crush wood at fittings; it will also excessively pre-load the wires. The completed wing should be flexible torsionally, but not sloppy. If one corner of the panel tends to lift when it is laying on saw-

Squaring the wing by tightening or loosening the wires to get exactly equal measurements *diagonally* between spar center-line points at projection of the compression member center-lines. Lateral distance between compression members is generally not the same so the measured distance is of no importance. Precise spar-to-spar distance must be precise. Internal wires must not be too tight; it is possible to warp a wing by excessive or uneven wire tension.

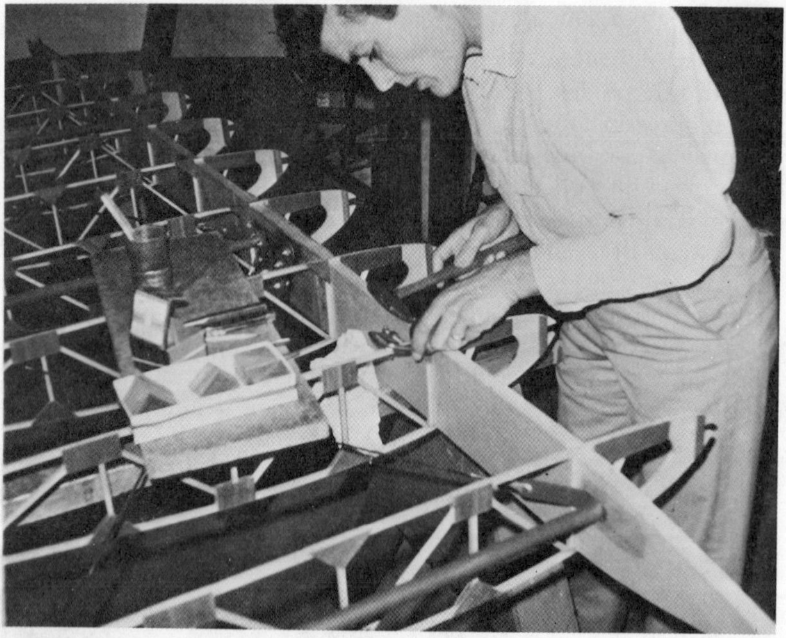

Ribs are glued and nailed to the spars after trammelling and the rib-to-spar trusses are set in. The gussets then go on. The small C-clamp holds each pair of gussets until enough nails are in each side to hold them to complete nailing. Bucking-bar in left hand backs up the work while driving the nails. A small drift-punch used with the hammer drives nails in too-close quarters.

horses, one or more bays of wires is either too tight or tightened unevenly. All should have the same “feel” of tension; use a tensiometer if you have one. All will have the same (musical) tonal pitch when strummed with the fingers, free of contact with the other wire or a rib. Be sure the tie-rod or wire is not twisted.

The threaded ends are next safetied by the appropriate means provided and some type of separator should be secured at their intersections to keep them from vibrating against each other.

After the structural fittings and wires are all secured the rib trusses which were left out are glued and nailed in their respective places on the spar faces and at the same time the cap-strips are glued and nailed to the spar edges (top can be done later after turning the wing over, if you prefer). The gussets for these junctures are added with the help of a small C-clamp to hold each pair until several nails hold them well for putting in the balance. A hand-held bucking bar is a necessity to back up the rib while setting the nails in all gussets while working on the wing. Longer nails are used to attach the cap-strip and trusses to the spar; I prefer No. 20 x ¾″ (two to each part: both sides and both faces of each spar). This system insures an absolutely tight glue joint between the ribs and spars. If the truss to the front face of the front spar was built into the rib in the jig, slip the rib slightly to one side and apply the glue to the place where it is to be on the spar, then reposition the rib and nail it in place and proceed with the others as above described. See photos illustrating this as used on a Stearman upper wing and also the smaller biplane lower panel.

For a small wing with ¼″ rib material and spars less than ¾″ thick, No. 20 x ½″ long nails should be used in place of the ¾″. Never use nails larger than No. 20 for rib work.

Following this assembly method, trammelling the wing before securing the ribs and then putting in the final trusses *after* trammelling, the wing is true and square throughout and the ribs are "built-in" without any "built-in" buckling or stress. With all ribs from the same jig the contour and fore and aft edges will all be uniform.

Most wings have a spruce strip fitted into a notch at the leading edge of the ribs to which the leading edge plywood or aluminum is nailed, and this should be put in next. The notches in the ribs may have to be trimmed or shimmed slightly for a good straight-line fit the full length. Put glue in the notches and slip the strip into place; nail at each rib with No. 18 nails which are long enough to go well into the rib nose-block, generally about 1¼″ long. The strip should extend forward of the ribs far enough (about ⅛″) to be hand-planed to the leading edge contour. This strip will strengthen the leading edge so the wing can be turned over on it, to work on the upper side, and preclude the forward part of the ribs being broken while the wing is worked on. Any intermediate "false" ribs would be put in place before the nailing strip is put in, of course. This span-wise piece adds to the strength of the entire leading edge by spreading the loads of minor bumps and bashes, and a plywood leading edge cover is much stronger than an aluminum one which is only nailed. That

Wing-walk support structure for the Hatz biplane. Top contour of ¼″ spruce root rib is planed down so the 3/16″ wing-walk will be flush with the remainder of the wing surface. It will not be carried forward of the front spar. Note smooth finish of spar roots.

serves as a form only. The center-section leading edge definitely should have glued and nailed plywood leading edge for the extra strength; the wings are literally hung from the center-section.

If the leading edge radius is too sharp for dry forming of plywood over it, wet the outer surface. Nail first along the leading edge, then up over each rib, then along the nailing strip (between ribs) on the spar edge, bottom side. Varnish thoroughly (2 coats) inside and on the inner surface of the cover piece (except where glue will go), then glue and nail on the second (top) cover. The leading edge will not be put on until after the root and tip ends have been built up however; that nailing piece lead our discussion somewhat out of sequence, but we do want that in place so the wing can be safely turned over so the rib and truss assembly can be finished up.

The root end may have a plain solid rib or a built-up box

assembly with reinforcing diagonal braces, etc. Be certain to make plenty of drain holes in the bottom surface of any boxed portion. If you are assembling a lower wing panel the wing-walk must be built in at this time. This must be of sufficient strength to be weight-bearing for a big man to stand and work on.

It can be strong without being too heavy by building the underlying ribs of solid plywood web (with lightening holes for the tierods to pass through), reinforced with vertical trusses or a standard rib with full plywood on one side instead of gussets. There must be sturdy span-wise pieces (approximately ½″ x ¾″ min.) supported by saddle-gussets glued and nailed at each rib in sufficient number between spars, and aft to the trailing edge spruce wedge-shaped piece (wide enough — about 2″ — to provide

Under side of aft wing-walk has been varnished and also the inner surface of the cover piece. Note drain holes in that and in the boxed root reinforcement which helps the diagonal braces prevent inward warping of the root. Seaplane grommets should be used over all drain holes, particularly where dirt and water will be thrown up by the wheels.

Wing-walk and root ends of spars after varnishing. Never cover spar ends.

adequate gluing area) where the wing-walk material will taper to fit the balance of the trailing edge metal. The bottom side of the area aft of the rear spar is usually also covered with 1/16″ plywood for added strength and also to prevent rocks being thrown through the fabric covering by the main wheels.

3/16″ plywood is recommended for the wing-walks. ⅛″ is too thin to hold up and ¼″ is too heavy; it is also too solid to be formed (wet) over the leading edge if you wish to carry it over the front spar to the leading edge. The 3/16″ plywood will require steaming to do this. If a form has been made for bending the rib capstrips it can be used also for bending the leading edge plywood for the wing-walk. No. 18 nails can be used for nailing down the wing-walk to the sub-structure, and smaller nails at the trailing edge area, along with blocks and clamps.

Diagonal braces of less than ⅜″ are too small to be of value at any place; ½″ square are better. Over ⅝″ becomes mere accumulation of weight. Corner-blocks should also be kept reasonably small.

Up until now, no mention has been made concerning the tip ends of the spars. We assume they have been cut to approximate shape according to drawings, but the final cut of the angle for bow attachment must be made at this time and it will be made to exactly fit the curve of the inner part of the bow made for it.

Wing sub-spar (to which aileron hinges attach), leading edge strip, roughly planed tip bow and diagonal braces, aileron well box reinforcing and nailing strips on front and sub-spar are in place. Clamps hold the basic hand-hold forms and spruce added at front corner to provide more gluing area for the tip "box". Corner-blocks yet to be fitted.

Basic hand-hold form in place. Corner-blocks yet to be put in.

Top, primary aileron structure together, the bow has been cut to separate it from the wing. Completed aileron structure is again hinged to the wing. Edges of gussets have been feathered with coarse file and sanded. Ready for varnish. Center, tip end from front quarter after varnishing. Papers on the floor to catch drips. Bottom, woodwork finished, the wing is ready for metal leading and trailing edge installation.

The *ends* of the bow must be surfaced to meet the side of the rib to which it is secured at the front and at the rear at the trailing edge. This may or may not be the same rib; in many wing plans, the trailing edge of the bow will extend around back several rib spaces farther than at the front, depending on how symmetrical or how eliptical the shape of the wing tip. If it tapers aft eliptically, attaching to several rib trailing edges, these will have to be modified (tapered more abruptly) to meet the section of the bow to make a clean contour, over-all. In any event, corner-blocks will be used at each juncture with ribs and after the bow has been hand-planed to shape on both top and bottom surfaces, gussets will be glued and nailed at these points, then they will be surfaced (feathered) for clean smooth edges fairing into the bow contour. The gussets will be of moderate size but large enough to cover the corner-blocks and strengthen the glued joints. If they are too big, they will later warp and look bad and the excess does no good.

If an aluminum tube tip bow is used, groove the spar tips to fit about 1/3 way around it to help keep it in correct position and also fair the fabric covering over it nicely. There will usually be a metal strap around it at both spar tips which is nailed, or perferably screwed, to secure it, and small tabs for attachment at the rib ends. Be sure to mount the tip bow at the angle specified; some follow rib center-line while some maintain a straight line with the upper wing surface or the lower wing surface. It could make a difference in the flight characteristics of the airplane.

Put drain holes (min. ¼″) in any bottom-side gussets where moisture could possibly collect.

All corner-blocks and other reinforcing, all remaining fittings, controls, cables and pulleys, airspeed lines and other tubing, electrical wiring, etc., must be installed and secured (and safe-tied) before putting on the leading edge plywood or aluminum. Necessary small clamps are best attached to spar with small wood screws (pre-drilled holes). Tubing and wires should be stood-off around, or attaching to, ribs to preclude any chafing. Use heavy-duty tie cord and shellac it after finishing the work; it will last longer. Remember: 2 coats of Marine Spar Varnish on the *inside* of all boxed sections, anyplace.

All metal fittings and the inside of all metal cover, such as the leading edge metal, should be sprayed with Zinc Chromate to prevent rust and corrosion. All controls must be lubricated before installation.

A plywood boxed leading edge, front quarter, tip provides both strength and a nicer contour than is obtainable with aluminum. Wetting the outer surface of the plywood helps to form it over the compound curve. If it is to extend beyond the outermost tip rib, it may be better to make a scarf joint over the tip rib and put it on in two pieces, both upper and lower side. Again: varnish inside surfaces. Spread the glue on all surfaces and nail it first along the straight-line nailing strips on the spar edge, then down the rib cap-strips, then along the wood bow which has been planed to shape. This part will take a lot of nails to get the plywood down firmly. Allow it to dry at least over-night before rasping it to shape, then sand smooth. Of course, the plywood box cannot be used with the aluminum tip bow.

A similar box-type enclosure is frequently used around the aft portion of the center-section (as the leading edge) for both added strength and better contour. Quarter-inch plywood formers are glued and nailed to the rear face of the rear spar, with added corner-blocks, and the plywood cover, as at the front, is glued and nailed to nailing strips on top and bottom edges of the spar, the edges of the plywood formers and the planed trailing-edge bow.

Hand-holds in the wing tips of low wing airplanes or biplanes are a great convenience in ground handling and also great fabric savers, because of damage to the cover if hand-holds are not provided. At least one is a must for a biplane, especially so if there is a center-section mounted fuel tank. These would be built in when finishing the tips and before boxing in the aft part of the center-section. They are well worth the small amount of additional work.

There are usually diagonal spruce braces from the outer radius of the wing tip bow to the inner faces of both spars; larger wings may have one or more in the rear quarter also. Gussets should reinforce these at each point of attachment.

Wingtip light attachment bases will be added and shaped and prepared for the mounting of the light after covering and doping is completed. The wire will be secured (with a loop of extra wire) to the tip rib with the end within reach after covering.

Finished trailing edge from the upper side. End corners bumped round to protect the fabric. Lug at aileron sub-spar is for interconnecting tube to upper wing aileron.

Trailing edge aluminum is held in position, drilled and riveted separately at each point with care to maintain overall length. Rivet heads on upper surface.

Check that all items are safetied before varnishing the wing. The trailing edge aluminum will be put on after varnishing, regardless of the method of attachment: nails, screws or soft rivets.

Varnish the wing completely before installing either hydraulic or electrical equipment such as for gear or flap mechanisms. Again check for safetying of every item. It is surprising what can be overlooked.

Another light sanding of the entire wing is advisable following the first coat of varnish, particularly all corners and edges of the ribs and gussets. There should be NO rough surfaces or edges any place. Brush out any globs of varnish on the lower side before they thicken. Turn the wing over to varnish the reverse side; it is the only way to work the varnish into every cranny and pocket. Observing all of these little details will give a prettier and longer lasting cover job.

New wood preservative finishes are coming on the market which may prove superior to Marine Spar Varnish for aircraft work but some are not compatible with the cover finishes (Butyrate Dope, etc.). I am generally inclined to let others experiment; but to date, Varathane looks good, although it is expensive.

Fresh varnish is thin enough for good penetration at room temperature. Warm it, if working in a cold shop, for *penetration of the first coat and spreading* of the second.

Use a good quality 2" paint brush, not too big to get into corners. Do not slop it on in excessive amounts which only runs to the low spots in undesirable thick blobs, but brush it out evenly with care to leave no spots of bare wood. If thick drops are allowed to set up, they make lumpy places under the fabric, and any irregularities under the fabric are always magnified on the finished surface. Plywood covered or boxed portions will generally absorb an additional coat of varnish to show a gloss.

The butt ends of the spars require additional varnish to seal the grain end. A wood sealer is good, or repeated coats of varnish for a good gloss finish of the cleanly sanded ends. They are *not* to be painted *or covered*, but open for close inspection for cracks or weather-checks. It is very unwise to allow such items to pass without repair. Smooth root ends can be kept clean and in good condition; they will not harbor dirt and moisture. If a "boot" type wing root attach fitting is used where the end of the spar can not be seen and inspected, it is even more important that the butt end be given extra protection. Provide some drainage from the fitting area for venting any moisture which could get into it. This is not a good arrangement, primarily for this reason.

It is not satisfactory to spray varnish on wood airframes because there is no way to get it into the numerous small, deep pockets where it is most needed, and there would be excessive amounts on surfaces from which it would just drain to low places. Hand-varnishing is tedious but it is the only way. A smooth and uniform gloss is desired. More is too much.

The accompanying photos illustrate the installation of leading and trailing edge aluminum. The trailing edge metal can be purchased preformed. The bent-in edge flanges are flattened with duck-bill pliers to be fitted to the ribs, wing-walk, etc. Tap it firmly into position, hold tight, drill, then set the soft aluminum rivet — one at a time. Before drilling, check the measurement from rear spar to the trailing edge for accuracy and uniformity. Rivet head goes on the upper surface. Cut rivet for no more than ⅛″ to be "set"; excess could buckle the rivet and split the rib tips.

It is difficult to hand-form the leading edge metal to correct rib contour, particularly the lower side, which has a more abrupt bend at the front leading edge. The upper section follows a more natural curve. With separate top and bottom sections, nail first the front leading edge, with about ⅝″ spacing, then up and back along the rib nose formers, starting in the middle of each piece and then toward each end of the piece. The metal *must* contact every rib continuously so there will be no bulges nor buckling. Nail spacing can be whatever is required here from ⅝″ to 1½″. Nail along the spar-edge nailing strips with about 1½″ spacing, then lightly break over the extending metal with a sheet metal hammer so it will not contact the covering fabric. Similarly bump down edges and corners of all leading and trailing edge metal. Cut-outs for spar fittings, airspeed lines, etc., must be anticipated and neatly cut before the metal is installed.

One-piece, wrap-around leading edge metal should be started by nailing it along the bottom edge of the spar then securing it around the ribs tightly and finish by nailing it along the upper spar edge. This is difficult at best, and particularly so if the metal

Starting the leading edge metal. Full wrap-around type would be started at bottom edge of front spar and finish along top edge.

Completed wing panel (lower left) for Hatz Biplane. All four the same except this one with wing-walk. Length: about 10'5" with 50" chord.

is old and stretched and bent as is generally the case if it is from an old wing.

Plywood leading edges do contribute greatly to the structural strength and for this reason it is definitely recommended for center-sections which support the entire wing assembly. It is also more resistant to dents and bends and helps maintain the leading edge contour.

So with this, we complete one wing and what has been learned will make the rest easier. Basically, the techniques are the same throughout the wood airframe structure. Every plane will present its own peculiar problems, but a little thought and ingenuity will solve them. Almost invariably, someone will say that it looks too pretty to cover. A job well done will continue to look pretty for many recovers. It will stay strong and resilient and you will love it!

9. Repair of Wood Airframes

Fuselage

There are very few old wood airframes left from WW I and the early 1920's, but there are some and they may be repairable. Wood is not condemned for age alone. If it has been stored in a dry, clean place, free of mice and tampering men, it can be safely cleaned up and restored to an airworthy condition. The most likely locations to expect deterioration are under fittings and in enclosed places where moisture could not be drained and ventilated, and in oil soaked areas where there had not been enough varnish to protect the wood. Many old planes had the longerons and other highly stressed members wrapped with fabric for reinforcement and if this has been dry it will have protected the wood over the years. If it has been oil or water soaked, it would have hastened the deterioration. Planes of this age are generally the rarely-found "basket-cases," and if one is fortunate enough to come across one in *any* condition, he can consider himself lucky, indeed.

Engine bearers of these old planes were usually of oak. Most of them carried inline up-right or inverted, or "V", or even "X" engines on parallel bearers having diagonal bracing. When we see these, then look at the huge old Liberty engines used in the big DH-4 and other craft, the common reaction is awe and respect for the courage of those who flew them. Fly them they did, all over the world, in and out of pasture "airports", for that is all there were. They would run out of gas, or oil, or water, or VFR weather and daylight, but not out of guts. They pioneered our aircraft industry both private and commercial. The only people who now laugh at the men and those machines are those who know nothing of either. And they have no knowledge of the wonderful qualities of the wood structures that did *not* fail them.

It was NOT the wood that would fail as years went on, but the old casein glue which gradually disintegrated. So if you find an old plane with wood wings or other parts, which was built prior to WW-II with the casein glue, you will probably have some rebuilding to do even though the wood may be in good condition. You would be safe in re-using the basic wood parts, but be prepared for a total tear-down and rebuild. A halfway job is a waste of time and materials.

A few remarks here in regard to what constitutes an "antique" airplane. The out-of-production planes are grouped loosely into the periods of time during which they were built and flown and for all practical purposes the Antique Airplane Association recognizes the pre-World War II aircraft in the various divisions of "antiques". Later aircraft are neo-classic, etc. Generally speaking, unless a particular and rather rare, pioneer, historical aircraft is of the older, composite type of construction in the more or less typical design of this colorful and often dramatic period of time, it is considered "modern" (mass production). Those earlier airplanes were actually hand-made. The fact many first World War planes were "mass" produced was due entirely to the circumstances; *many hands* were employed to build them. There were no computerized machines operated by punch-cards. Of course, they were not custom finished either and when we see a very old machine still "as original", in a museum or found in an old barn some place, this is very obvious. The beautiful re-built WW-I planes at air shows are far better in nearly every way from the way they were originally turned out. They were built in a hurry to fight a war or to be used in training and the one was as hard on machines as the other. That is WHY they have been so rare long years before now. Life was about as hazardous for the WW-II British Mosquito Bomber and the thousands of wood-winged training planes still fresh in the memories of our own generation, so we are glad to see some of these being put back into first class flying condition and used for sport flying now.

The big dividing line between wood construction as compared to metal is that wood is hand-worked while metal lends itself best to mass production methods. The new light weight, high stress metals in use today brings entirely new construction processes and concept of flight; it has put men on the moon. It is far beyond our personal experience but probably not that of the "home-builder" of the future, and that will probably come faster than we think in view of the rapid changes of recent years. This has surely been a fascinating period of time to live in. Maybe we can't all keep up, but it is fun to watch. We regret the destructive uses of so many potentially good things. When man forsakes his ideals the results are disastrous. I believe that when a person *makes* something, the benefits extend beyond ego-satisfaction, especially in building or re-building an airplane in which he will fly. Flight itself gives man a wider view of the world and tends to give him a wider respect for its grandeur and complexity.

The mechanic's "bible" is the FAA publication: *AC 43.13-1* and *43.13-2* which are available from the U. S. Government Printing Office and well worth their small cost. In fact, *43.13-1* is a *must* for anyone working on aircraft. For our present purposes the applicable portion is Chapter I, pages 1 through 24, "Aircraft Wood Structures." Refer to this for all wood building or repair. If you have a problem not described, *ask* your local FAA General Aviation Maintenance Inspector for advice. This little book is intended to help with the many little "how to — " go about things, whether old or new. It is not a substitute for accepted requirements. Other people may have better ways of doing certain parts of the job; this is the result of practical experience working with both old and new planes with commonly used and simple equipment, as listed under "Shop & Equipment." *All* aircraft work demands patience. If you do not have it, forget it. But if you know and love airplanes and like working with tools you can learn the rest.

This little dissertation is inserted here because very often, "repair" applies to the older, ATC'd type aircraft involving more restrictions concerning just what work the owner is permitted to do if he is not a licensed mechanic. He is required to work under the "Direct Supervision" of an appropriately rated and licensed mechanic — and to save you much trouble, you'd better believe it! There will be questions later. We advise anyone who contemplates any major repairs to any ATC'd airplane to make satisfactory arrangements with the A & P mechanic and your local GADO *before* starting to work. There is too frequently a sour note: many shoddy jobs are done just to get a plane by to sell. This is irresponsible and contemptible and could well be considered criminal. Do it right or don't do it. (Buyers, *Beware!*)

So on with the inspection of an old airframe which you think is worth saving. After cleaning off all surface dirt, remove all metal work, leaving the wood bare. All metal clamps and fittings of every kind must also be cleaned thoroughly, inspected — and by "inspected", I mean *looking for trouble;* especially little cracks at bends and welds, elongated holes, edge distances remaining around holes, fittings worn thin, deterioration from rust or corrosion, chafing, etc. This inspection must be done by someone familiar with the many faults to be *looked for.* An experienced eye will spot defects unnoticed by a novice which can be very critical to the strength and action of a part. All paint must be removed and steel parts sand-blasted, if possible, or otherwise hand-sanded to bare metal. They should all (steel or aluminum) be zinc chromated as soon as possible following inspection and repair to protect them from rust or corrosion, then put them away until needed for reassembly.

It is not adviseable to use paint or varnish removers on wood airframes because many of these products contain chemicals which might penetrate the wood and not be removeable; a certain amount is bound to soak into the wood and be absorbed, with unknown results later. I prefer to use fresh Butyrate Dope *Thinner* for all such cleaning purposes. It dissolves nearly everything so that old paint and varnish can be scraped off clean with a knife or scraper (do not gouge the wood), and it will leave no harmful residue. Allow the wood to dry then sand it completely. Do not remove any more wood than can be avoided but get it all as clean as possible, particularly where there will be new glue joints. Glue will not properly bond to other than bare wood.

If there are gussets at the joints of truss members and longerons or plywood covered in all or part, remove loose or warped or broken plywood and all nails and screws. Scrape and sand away all old glue. Old casein is by now usually dry and powdery and can be sanded off easily. New glue requires filing and scraping with a sharp tool. It is far more satisfactory to replace warped and distorted sections, gussets and reinforcements, then try to splice and repair them; this particularly is true of skinned areas. It depends on the age and condition of the plane. If it is just an area of limited damage, for example, the skinned tail surface of a Beech Staggerwing, a local minor repair is obviously in order, as per *43.13-1.* No one would expect you to strip down an entire component for minor repair. But where an entire rebuild is involved, it is a lot easier, quicker and better to rebuild a complete unit "as per original" as closely as possible, which of course means the same kinds of materials and method of construction. Original pieces should be used as patterns if drawings are not available. Often, there will be vast areas missing, so unless you have access to another similar part, it will be necessary to get drawings from an individual or FAA records.

The Regional offices of the FAA keep copies of factory drawings of all aircraft ever produced by manufacturers in those Regions and it is possible to inspect these drawings or have copies made (with the written permission of the person or company currently holding the manufacturing rights). Some libraries have invaluable copies of books with very accurate and informative descriptions and drawings of aircraft built since the earliest of flying machines. Libraries of engineering schools would be the most likely starting place to look, if you have entrance. Many individuals collect such information and can often be of immense help to a restorer trying to get an old plane back into flying condition. An advertisement requesting the specific information may be productive.

Professional advice is urged and recommended regarding replacement of old metal parts, such as 4130 steel in place of the 1020. Unusable cast fittings (or missing ones) are a big and most often expensive problem. All substitutions of such items are closely scrutinized and approved as "equal or better". Struts, engine mounts, landing gear, wires and all fittings are just as vital on an *all-wood* airplane as any other part of it. This should be obvious but it is sometimes overlooked.

Rust and corrosion weaken not only the parts they are on but also those adjacent to them, including the wood; moisture causes it and then it holds moisture in return. Moisture is the one thing we most want to avoid and get rid of in any part of an airframe. The condition of any aircraft structure must be considered in its entirety.

When everything has been cleaned completely the vital points must have special attention: fitting areas of the landing gear, wings, tail surfaces and struts. In a wood airframe these items are seldom directly connected, but through fittings which are often rather complicated. The fittings bolt to the airframe, so it is at these points that very close attention be given to look for deterioration of any kind: cracking, checking, compression of wood due to over-tightened bolts or elongation of bolt holes.

Depending on original hole size and the dimensions of the member, it *might* be possible to go to the next larger size, but get advice from someone well qualified before doing that. If a hole is enlarged, to correct an out-of-round condition it is best to bush it so the original bolt size can be maintained, and the *original hole center* must be held, for proper alignment of the parts.

If wood is badly checked or cracked at bolt holes, a new section should be spliced in or the piece replaced entirely. Scarfing in new sections requires reinforcing plates which could seriously interfere with other members or fittings. If much of such repairs appear to be needed it would be more satisfactory in every way to use the old frame as a pattern only and re-use the old fittings if they are in good shape.

The tailpost area is always subject to the collection of dirt and moisture as well as the pounding and shaking of landings and take-offs. Most of the dirt and moisture from the front end on back accumulates there eventually and there is seldom adequate provision for drainage and ventilation. Dirt and bugs will often clog drain holes in the fabric. So when you build or re-build *any* fuselage, provide access for both cleaning out and draining the lower and aft part of it. A vacuum cleaner with a long extension is fine for removing much unwanted dirt and debris from any fuselage. Most planes have removable fairings or at least inspection holes through which a vacuum can be manipulated.

If the old fuselage frame is basically good, with only a few broken or missing members, by all means, save it.

It is a simple matter to replace loose corner-blocks and reinforcing strips, as per original. Sand away old glue and dirt, cut new pieces to precise fit and nail in the new ones using No. 20 or No. 18 nails of the appropriate length. Use the bucking bar to back up the work when renailing on the strips; the magnetized end of the tack hammer leaves the one hand free to hold the bucking bar. Thin parts *must* be backed firmly for driving nails. Use short nails which will *not* go clear through the base piece of wood; that is a good way to split the wood. Staples are not desirable. They require backing just as much as nails and they do rust, unless of stainless steel. Nails are as easily removed with the small diagonals and their heads are flush with the surface while staples are not. We build and restore with the idea of its lasting a long time, but is well to keep in mind the repairability factors because accidents do happen and there is always a certain amount of weather deterioration with any airplane; the all-metal ones have their corrosion troubles.

The fuselage structure, new or old, needs plenty of Marine Spar varnish well brushed into every corner and pocket (be certain that *all* dirt has been cleaned off first — don't varnish over dirt) It helps a lot to be able to rotate the fuselage completely so you can see better and be able to work right side up; brushing varnish (or dope, when covering) upside-down is a mess. Apply a minimum of two coats over all bare wood. If it has been only a clean-up job, one coat will probably be enough. An over-all gloss should show but there should not be excessively thick varnish any place. Make certain that all drain holes in gussets, etc., are clear and open. Even old wood should look fresh and clean; new wood will of course be lighter. The old wood has a mellow tone and smoothness with a charm all its own, and it will not warp any more than it has unless for some very *unnecessary* reason. A properly done fabric job will not cause warping, particularly of well-seasoned wood. (See *Aircraft Dope & Fabric* by Spencer.)

During re-assembly, use wide-area washers (only) under all nuts or bolt heads to increase the bearing surface and prevent crushing of the wood. Do not over-tighten nuts and bolts. Depressions in the wood (compression) caused by over-tightening may be cause for rejection.

10. Repair of Wings

Except for obvious structural damage to a wing such as a cracked or broken spar, broken ribs or tip bow, repair of old wings is an exercise in judgement based on experience and it is advisable to have someone look at them who does have that experience before optimistically starting to work. Many times, there is a hairline difference between repairing, or simply building a new wing, utilizing the old (reconditioned) metal parts. What to save; what to replace? Maybe change one bad spar and save the rest? Splice a spar or replace it with a new one? Splice-in rib cap-strips or make a new set of ribs? Splice-in a wing-walk section or make an all-new one? The problems that can be involved are endless. Let's consider some of these things and weigh the odds in regards to time, money and serviceability.

Begin with a typical case of obvious damage. Say the plane has been ground-looped; it dragged a wing and the rear spar has been fractured. In a high wing monoplane it would probably be cracked out at the tip. The bow would be bent and scraped if metal; scraped and probably torn loose from the spar, if wood, with the diagonal braces broken loose and most likely the tip ribs also loose and broken. The sub-spar which carries the aileron and the aileron itself, may also be cracked, or scraped. A biplane would have similar damage but it could possibly have carried through to the upper panel via the N-struts. One time, I knew of an old biplane in which every spar — all four wing panels — were broken by this transmission of the shock; not too uncommon in the biplanes built in the 1920's and '30's. They were built to fly, not bash. This is good enough reason for a thorough inspection of any plane following a ground-loop type mishap.

As *Manual 43.13-1* says, there can be NO splices at a fitting (compression member or control, etc.); this includes the splice reinforcing plates, so if the spar crack or fracture is outboard of the tip compression member, a splice out there would present no problem. It would be necessary to remove the compression member and one or two ribs to allow room to saw the long (min. ten times the thickness of the spar) diagonal splice cut, and have the required over-lap distance for the reinforcing plates. Also at these plates, the rib vertical trusses which glue and nail to the spar faces must be relocated to accomodate the plates. If the ribs are solid plywood the spar cut-out must be widened. It is quite obvious that any spar splice further inboard in any wing will involve a lot of work and it must be left to the judgement of a good mechanic to decide to splice or replace. Naturally, the more complicated the spar construction, the more difficult it is to make a satisfactory splice. Unless the wing is in excellent shape, I recommend a splice for tip damage but spar change for inboard damage; it is easier.

Some relatively modern airplanes have all-wood wings and tail surfaces, for example, the Beech Staggerwing series, the Cessna Airmasters, the Fairchild 22's and 24's, to name some of the best known types frequently seen in use all over the country. The Airmaster has a full cantilever wing with plain spars, truss type ribs and plywood skin over the leading edge (the main spar is set about 1/3 of chord from leading edge) and the empenage is constructed similarly. The Fairchild wing is externally braced with lift struts and is of conventional construction; the spars are of "I"-beam section. The Staggerwing wings are of conventional construction with plain solid spars and closely spaced, truss type ribs. Their tips are plywood skinned inboard almost to the inter-plane struts; some models have plywood leading edge and some aluminum. This is a high performance airplane with maximum speed approaching the 200 mph range, so the rib stitching of the fabric cover is 1″ spacing through-out. All of these planes carry fuel tanks in the wings. All of them perform comparably and favorably to the later models built by the same companies, with the exception of Fairchild, which stopped production of light aircraft after the WW-II PT-19 and 26 series. They show very well indeed against their later, younger brothers with their flat engines and shorter props and all-metal construction. But there is no doubt that the all-metal plane is mass produced much faster and cheaper.

With reasonably good care the wood-winged plane will last as long as any other and it takes a pretty rough bash to seriously damage a wing. The main thing to watch for is adequate drainage of moisture in all wood parts. Be certain there are drain holes where they should be — and keep them *open.* Seaplane-type drain grommets are far superior to the plain hole type because the holes in them are larger and because of the active venturi action which vents the interior so much better.

The fuel tank installation should be inspected rather often to be sure it is dry and clean externally and not retaining either moisture or gasoline. Either would cause deterioration of the wood structure, and dirt *anywhere* holds moisture; so the cleaner the better.

Watch for abrasion and punctures in plywood-skinned tail surfaces caused by flying stones when the plane is operated on dirt or gravel areas. Such damage is frequent and should receive immediate attention. *AC 43.13-1* describes the methods of repair for plywood skin damage. It is not difficult, but it does take patience to do it properly and neatly. The first thing is to sand off all varnish or sealer in the area so as to get securely bonded glue joints. And do not forget to varnish the *inside* of the patch before gluing it in place.

Whatever the style of wing construction, repair must begin with disassembly. If it is skinned, set it on a sturdy pair of saw-horses with horses under (not between) ribs; nail-hole lines will indicate where the nailing strips had been. Do the same if fabric is still on any wing. If there is no cover on an unskinned wing, set the horses between ribs, clear of fittings. Remove ailerons and flaps (if any), then any fuel tanks and control mechanisms.

To uncover, cut fabric along the trailing edge and around the tip, root, and any place it is secured in aileron well, etc. Cut it chord-wise in two or three places to make it easier to handle instead of trying to take it off in a full span-wise piece, especially if the wing is big. Pull up and forward, cutting the rib stitches as you go. The lower half will fall away fairly free except over the saw-horses; the wing must be lifted to pull the old fabric clear here. There will be some places that will have to be cut and pulled loose; be careful not to damage under-lying structure as you want to save as much as possible intact, just for patterns, if nothing else. It is adviseable to save the fabric as intact as possible too, as a guide when making the new cover.

If the structure is not very familiar in all details it must be closely studied and sketches made for duplication at re-assembly, or if there is a pair, hold the better one 'til last, to use for reference throughout the time its mate is being worked on, then the new one will be the guide. Duplicate parts can be made for the pair (or set) as you go. If the same portions are gone from them both (a pair of *both* upper and *both* lower panels of a set for a biplane) it will be necessary to get drawings or another panel of like kind that is in reasonably good condition. The accompanying photos of the wings for the D. H.-60M Gypsy Moth wings illustrate what I mean by this. One upper and one lower panel were shipped to me to serve as patterns, and a full set of metal parts was supplied after much searching. Mr Kelly was able to furnish copies of deHavilland drawings for rib construction and other details. New aluminum trailing edge metal was put on the new wings (no leading edge cover). A completely new set of wings was built.

The original D.H. -60G Gypsy Moth was built in 1927, all-wood construction; some of these airplanes set very respectable long distance records, all around the world. Later models were built with steel tube of the same dimensions (fuselage), but wings and tail remained wood throughout the period of manufacture in England and in the U. S. A., where they were built under license from deHavilland with the Wright-built engine installed. The British-built planes were used in great numbers all over the world for a great diversity of purposes from training, sport flying, to crop-dusting. The leading edge flaps are free-floating and effective. The inter-plane struts are wood set in socket-type fittings. In more recent years they were succeeded by the swept-wing Tiger Moth which has more horsepower and a few other minor improvements. These were the Canadian Primary Trainers of WW-II and quite a few have recently been imported from Canada and Australia as sport planes. Their construction is basically the same as the Moth. Tiger Moths were never built in the United States. All of these planes have deeply routed spars and they do get expensive to replace.

Your wing may need only repair of minor damage. This is a relative thing to a person accustomed to working on them, but the FAA does have specific ideas about major or minor repair of anything, so do consult your local FAA Maintenance Inspector on this point and get his advice. We can not authorize anyone to do anything, but advise how to go about it. One time, I received a letter from some man wanting me to write him a letter about how to rebuild his wing! Some twelve or fifteen years later, this is that "letter".

It is a lot of work to change a spar, but if the damage is any place other than out at the tip, I think it is preferable to splicing.

"Before," lower left panel, one of the pair received as patterns for building a full set of new Gypsy Moth wings, supplemented by DeHavilland drawings. Ailerons on lower panels only.

— and "After," new lower right D.H. 60-M panel. Aluminum tubing wing tip bow.

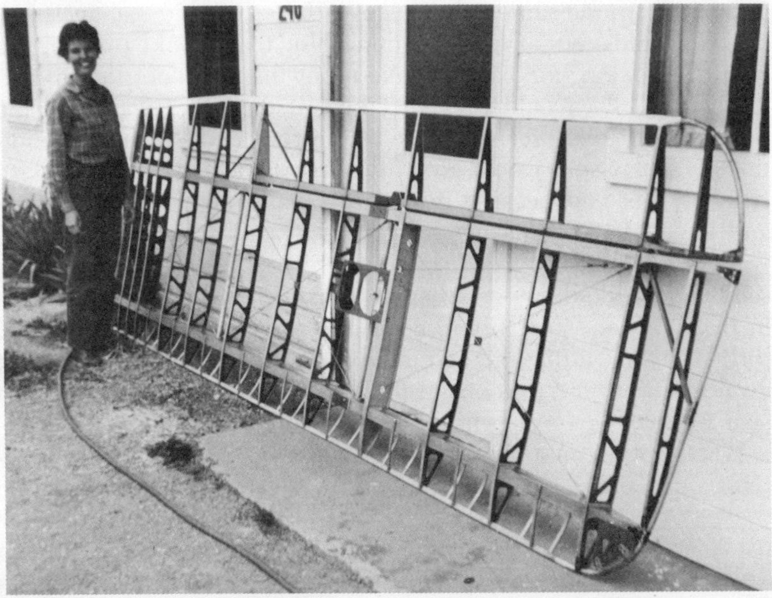

There is *nothing wrong* with a well spliced spar, but it must be done well and according to *AC 43.13-1,* which shows the proper methods for various representative types of spars. The important point in making any splice cut is that it be clean (straight) and a perfect match with the new piece. The mating surfaces must be precise and the total absolutely straight and true, both horizontally and vertically and with both faces also straight. It is easier said than done.

Cut the spar off in the wing first, then the new piece to match it. Use a hand-plane to even-up both surfaces, checking with a metal straight edge until both are trued. The new section should be cut extra long to allow for working over the splice cut for perfect fit; the tip end can be trimmed later. The edges should be bevelled before making the splice cut. Bolt holes for fittings in the new part will have to be drilled after the splice is completed, as per original. It requires a lot of care to get them drilled exactly 90°, straight (both ways) with the spar set vertically in the wing. A fitting plate can be clamped in place as template but be careful not to elongate any holes in it with the drill.

To remove a damaged spar, take out all metal work, starting with the leading and trailing edge, any wires and tubing, controls, etc. If ribs are to be saved and the spar re-used, carefully remove all nails from the rib trusses at the spars, with the small diagonal cutters. Loosen the trusses with the point of a knife, prying gently. In the same way, remove the spar-edge nailing strips, both top and bottom, and anything else that would prevent the spar from being worked out. Take it out the root end; the tip and smaller tip ribs would prevent taking it the other way, and the heaviest spar reinforcements are at the root end, so the re-conditioned or new spar would go in, tip first. It takes a lot of work to free the root end because most upper panels have a boxed area and lowers have the wing-walk. It will be necessary to turn the wing at least once to free everything. Since the wing will be reassembled upside-down it is best to make the spar removal (final) in that position. Very often, getting a damaged spar out is more difficult than putting the new one in. It may be necessary to cut away some jagged portions in the area of a break, and often, the spar will have a very "bent" out-board part.

When putting in a new spar the process is reversed and everything replaced as per original. New or cleaned up corner-blocks, truss pieces and gussets will be used to reassemble the wing which will be accomplished as in the procedure described for new wings. A new spar will have been made up as per the original, with the bevel cut (planed), reinforcing plates glued in place, holes drilled for fittings, etc. Do not overlook the two coats of varnish at all fitting areas *before* assembly.

Sometimes old spars will have shrunk a slight fraction of an inch, enough to show daylight at rib trusses and cap-strips. Carefully remove nails from this loose part, and the gussets and the loose truss piece. Clean away the old glue and set in a new vertical truss with a new pair of gussets. Shim the space between spar edge and cap-strip, re-glue and nail (No. 20 size). Be certain the ribs are re-set square with the spars. This minor trouble is not uncommon and is easily fixed. There will usually be loose gussets at other places too, to be replaced with new.

In cases where the old cap-strips are pretty thin and flimsy, but the ribs are still sound and have good contour, glue and nail small corner-blocks alongside the cap-strips at all spar edges where the ribs cross the spars instead of trying to re-glue and nail them. Pulling the old nails and setting new ones could easily split them. In the first instance (above paragraph) to re-glue and nail down the cap-strip without shimming would result in a depression of the rib contour where it passes over the spar. Keep the dimensions of spar edge corner-blocks less than the cap-strip so they will be beneath the rib cap-strip level and not be in contact with the wing cover fabric.

Old wing tips bows are frequently warped and may have laminations at the aft position badly separated from weathering, again, from lack of good drainage and ventilation. Gussets at spar ends and rib connections will have warped and become loose, and often too, the corner-blocks which reinforce these junctures. The best repair is to replace them all with new, as per original. The bow laminations can often be re-glued with the use of clamps and blocks, taking care to maintain the proper shape and set of the bow. Sometimes a badly warped section of a bow can have a replacement piece spliced in, using an added plywood reinforcing piece past both ends of the area; the outside will be finished to contour, of course. The tip diagonal braces most often have to be replaced also, being often warped or split at the ends. Increasing the dimensions of replacement reinforcing pieces or gussets "to make them last longer" is not wise. Added weight *does* accumulate and just making parts bigger is no reason for them to *last* longer. Durability depends on the quality of construction and finish and on good maintenance.

In many instances the decision to patch up or to completely rebuild can be almost a toss-up. At one time I had a pair of lower wing panels to repair for a Stearman and they were in pretty rough shape; repairable, but a lot of work. The spars were OK but needed to be cleaned and reinforcing plates replaced, etc. Ribs were warped and loose and the wing-walks of both were tired, with a lot of warped and loose stuff underneath. Both, just typical old Stearman wings to be made airworthy again. The first one, I elected to "fix up" and nearly three weeks of full-time work went into the repairing of all of the above items, plus the tip bow, reworking the aluminum leading and trailing edge metal, etc., etc. It was sound and in good condition when it was finished but it had taken a long time and proved without a doubt that it was the hard way. The second one was stripped right down to the bare spars at the beginning and they were completely cleaned and the reinforcing plates fixed or replaced with new ones (much easier to do than while in the assembled wing). A new set of ribs was built, which took *less* time than scarfing and splicing warped or broken old ones in the wing; a new root and wing-walk was built in; the old bow was repaired and replaced and all new braces and supports and corner-blocks and gussets and front, wing tip quarter box built in (this had to be done on No. 1 too, of course) So No. 2 was like new and it had taken no more time and had been a lot easier. The final cost to the owner was the same and you know which he liked better.

Take care in all disassembly. Quite often, nails holding the leading edge aluminum will be very rusty and will not lift out, their heads will tear through the aluminum instead. In this case, just be as careful as possible not to allow the metal to tear at these holes. Work along the front edge, lifting gently with a thin screwdriver and pulling nails which do lift enough to get hold of them with the little dikes. Trailing edge metal secured with rivets must be removed by drilling out the rivets and try to avoid enlarging the holes in the metal here, also. After the metal is off, go back and pull *all* remaining nails; if you don't, you will be sorry. They are hard on both fingers and tools.

Total rebuild of an Ag. Stearman wing; all new wood except front spar and repaired tip bow. All metal fittings cleaned, inspected and zinc chromated.

Assembling the rebuilt wing. Ribs loosely in place until fittings and wires are installed and the wing trammelled, as per Chapter 8. Full new wing-walk to be built into far end, along with the root structure. A month's full time work to complete a lower panel such as this, and two weeks for an all-new upper. Stearman center-section also requires a month.

Take the time to bump out nail impressions and other irregularities in each piece of metal to be re-used. Clean them well with sand-paper or emery paper and thinner (to remove old paint or varnish and just plain dirt) and spray them with zinc chromate primer. They will be ready to install when the repaired wing is ready. Torn sections must be replaced with new metal of the same type and thickness. There must be no rough or protruding edges; all are to be "broken" over slightly before installation. All edges are to be taped with masking tape or an adhesive cloth tape before re-covering, to protect the fabric.

All metal parts should be thoroughly cleaned, inspected, overhauled as necessary, zinc chromate sprayed, and lubricated before re-assembly. Carefully check all bushings and bearings and the operation of all control mechanisms for smooth and correct action after re-assembly. Aileron hinge brackets must be in alignment, and of course the same is true for flaps. Many old planes have plain bushings in hinges and they sometimes wear out totally, like *gone.* Ball-bearings can and do also dry out and wear out, so that the bolt turns within the center bushing rather than the bearings rotating in the race. Bolts which show wear should be replaced, and *do* use the correct size, and avoid a stack of washers as spacers! There are established limits applying to this. In regard to bolts, this special note: Stearman wings were built with *all aluminum bolts except* the ½" bolts in both the front and rear spars which hold the flying and landing wire lugs at the interplane fittings (and wing to fuselage attach bolts). To arbitrarily replace these aluminum bolts with steel ones adds *pounds* and to no good purpose. Replace *bent* bolts, yes; but if the aluminum bolts are still in good condition, re-use them — and use *aluminum* nuts with them. It helps to put a small drop of oil on the threads and it will help to do the same also when removing the nuts. Sometimes they are old and very dry and sieze, and even have to be sawed or drilled to get them out. But odd as it may sound, I have found extremely few sheared aluminum bolts in all the years of rebuilding these wings; they have been bent — zig-zag and then some — but very rarely sheared. (Steel bolts bend too.) They may be corroded; if so, replace them. The aluminum plate-type fittings are often corroded too — and on the *inside* against the wood. This usually involves an intergranular type of corrosion and is due cause for rejection of the part. Stearman wing fittings ARE available on the market (to date) and there is no excuse to put together such a fine airplane with parts which could easily fail in flight. The forged strut fittings must be straight and in good condition, free of cuts or corrosian.

Loose spar reinforcing plates at fitting areas, especially at the root end at the wing-to-fuselage attachment, are a common problem in old wings that have not been well protected from weather. If there is no cracking or checking in the spar root or the plywood, just clean out the separation as well as possible (fittings must be removed), reglue and press with blocks and clamps. Do not just nail the plates to the spar; this does not give enough nor even-enough pressure for a good bond and the excess of nails is very undesireable. Allow at least twelve hours setting time, then sand off the surface and edges, revarnish and replace the cleaned up and chromated fittings.

Sometimes it is necessary to replace the reinforcing plates (in the wing or with the spars removed), if they are very loose, damaged, or badly deteriorated, or if deterioration of the spar itself is suspected. Pry off *only* the loose portions that come free easily. Try to determine which way the grain of the spar is running before further lifting the plate or cutting into it close to the spar. If it runs into the spar, work the other direction while planing off the old plate. Remove all nails as you can get to them; they are hard on cutting tools. Work very carefully down to the bare spar wood; do not gouge into it. Sand with a backing-block until it is clean and perfectly even; check it with a metal straight-edge. Brush or blow off all saw-dust. Pre-drill the new reinforcing plates (a pair together) using an appropriate fitting as a template (clamp fitting and pair of plates together so none can shift on the drill-press). All work must be held securely for drilling, for precise hole-centers and roundness. Always use the correct drill size; over-size holes are not acceptable. Fittings must be in correct alignment. Many times, these holes will be bushed, and there is no room for error.

When ready to glue on the new reinforcing plates on an old (or pre-drilled) spar, first locate them by inserting at least three of the bolts — or two, if that is all there are at that location — and hold the plates in place by setting in a No. 20 nail (½″-¾″ max.) at each corner and perhaps a couple in between if it is a long plate; that is sufficient to hold them until they are clamped. Use enough clamps for even pressure; the thicker the blocks (approximately an inch) the better for uniform distribution of pressure. Of course the blocks are used on *both* sides, always, and the plates are put on in pairs. The locating bolts will be backed out (*not driven* either in or out — *ever*; the surface of the wood on the "out"-side can be torn) as soon as the locating nails are in place.

Except at the butt end of spars, the ends of all plates should be chamfered with the hand-plane, and sanded. The thicker the plate, the larger the area of the chamfer, varying from approximately ⅛" to ½". This is to avoid concentration of stresses where the reinforcement ends abruptly, which is similar to putting a radius in bends of metal or tapering off reinforcements in metal structures. Rounded curves, corners and edges are always better.

As with a new spar, the butt ends are to be cleanly sanded and varnished. *Do not paint nor cover them.* Do not *add* nails to root-end rib or boxed cap-strip. Remove old ones, clean up, re-glue and re-nail with no larger than No. 20 x ¾".

We see all kinds of "fixes" for deterioration of ribs at the lower edge of the rear spars at aileron wells, all brought on by poor maintenance and lack of drainage and ventilation. Many of these repairs are inferior to the original, which holds up quite well with proper care. Wrapping formed aluminum around the whole thing is *not* the best way to cure the problem. For one thing, it adds weight and for another it only helps to box in the dirt and moisture we want to get rid of. Another objection is the difficulty in getting any fabric to hold to the aluminum in the concave surface so it will not interfere with the aileron (or similarly, the flap) action. Yes, fabric *can* pull away from wood structure in the aileron well, but not if it is done right and properly secured. Aluminum covering in there, over wood structure, offers no attachment for the fabric.

Another cause of trouble in this problem area is that people so often just drive in bigger nails to hold the cap-strips instead of repairing them correctly, thereby splitting them and promoting even more deterioration. The result is dry-rot (which is moisture deterioration) of the lower edge of the rear spar as well as the lower, rear part of the cap-strips. So eventually, there is not only rib repair but also a spar change in order to save the wing.

Splicing lower rib cap-strips there is not as difficult as toward the more curved forward portion. Refer to *AC 43.13-1* for rib splices. Cut the scarf with a small sharp back-saw. The matching cut in the new piece will be done on the band-saw, with final mating surface finished on the disk-sander. There must be exact, straight alignment, with the opposite end also fitted before gluing the scarf splice. Use small C-clamps (2) to hold and press a pair of small blocks on each side for the setting time (12 hrs. min.). Splices will be made clear of spars and elongated gussets set on each side; these are glued and nailed, as are other rib gussets.

There have been cases where *all* wing ribs have been broken at the leading edge and the owner wanted them spliced. I did. But a new set of ribs would have been easier and better. This can happen when a plane flips up-side-down (biplane or high wing types). However, if the fabric and the rest of the wing is in very good condition, it is sensible to splice them all. The condition of the whole panel must always be considered. This is true of repair in any part of any airplane.

Ribs in a fully skinned wing usually last a long time because they are well protected, and if the wing has been given good care while on the airplane or while in storage the whole structure will survive many re-covers. It is dirt and dampness that gets to any wood assembly and there are certain parts of the world where the normal humidity is so high that any wood airplane suffers seriously. Excessively dry country will allow a certain amount of detrimental shrinkage and subsequent looseness of fittings, so it is well to check them at frequent intervals. To go from dry to humid climate would cause the wood to absorb a certain amount of moisture and perhaps swell just enough to make bolts and nuts overtight. So when moving an airplane from one very different type of climate to another, give this special consideration and watch for these points at inspection time.

Wood is nice to work with. So if you are one of the lucky few to have an old airplane, all or part wood, "Keep Them Flying" and enjoy it. If you are building a brand new plane, don't let anyone tell you it is "obsolete".

Trees will never be obsolete, nor will the sky, nor the spirit of those who love both and are free to fly.